Alfred Binet

Animal magnetism

Alfred Binet

Animal magnetism

ISBN/EAN: 9783337228743

Printed in Europe, USA, Canada, Australia, Japan

Cover: Foto ©berggeist007 / pixelio.de

More available books at **www.hansebooks.com**

THE INTERNATIONAL SCIENTIFIC SERIES

ANIMAL MAGNETISM

BY

ALFRED BINET

AND

CHARLES FÉRÉ

ASSISTANT PHYSICIAN AT THE SALPÊTRIÈRE

NEW YORK
D. APPLETON AND COMPANY
1889

PREFACE.

WE think it well to state that this work was written in the environment of the Salpêtrière. By this we not only mean that our descriptions apply to facts observed in that hospital, but also that our personal observations were made in accordance with the method inaugurated by M. Charcot, the chief of the school of the Salpêtrière, that is, in accordance with the experimental method which is illustrated by clinical science. While relying on the observation of spontaneous facts, we have strengthened these facts by experiments.

It would at present be premature to write a didactic treatise on animal magnetism and hypnotism. This work only aims at giving an account of special researches which, notwithstanding their number and variety, will not justify general conclusions on the question. After receiving this warning, the reader will not be surprised to meet with occasional breaches of continuity, which are, however, more apparent than real, and which are due to our resolution not to speak of experiments which we have not verified for ourselves. Completeness is good, but it is still better to assert nothing of which we are not assured.

B. AND F.

CONTENTS.

ANIMAL MAGNETISM.

CHAPTER I.

ANIMAL MAGNETISM IN ITS BEGINNINGS—MESMER AND PUYSÉGUR.

WE propose to discuss a question as old as the world itself; which about a century ago was admitted into the sphere of scientific discussion; which, although constantly rejected and disclaimed by learned bodies, has always reappeared, and is still in process of evolution, notwithstanding the importance of the results already achieved. In retracing the history of animal magnetism we shall endeavour to explain the causes of these vicissitudes of fortune, and to indicate what instruction may be derived from them. As we proceed with our subject, the truth will become more evident that it was owing to a lack of method that animal magnetism was not admitted at an earlier date to take its place in science.

It concerns scholars to trace the course of animal magnetism through the ages, and to seek for its remote beginnings in the customs of ancient peoples. We refrain from such historic studies, for which we are incompetent, and propose merely to sum up the conclusions of science with respect to animal magnetism, and consequently only to speak of its history so far as this history has left its traces on the present state of the question.[*] From this point of view, it is unnecessary to go back to an earlier period than that of Mesmer and of his immediate predecessors.

Mesmerism is connected with a tradition which had its origin towards the middle of the sixteenth century, a tradition which, as the name of animal magnetism implies, not invented by Mesmer, ascribed to man the power of exercising on his fellows an action analogous to that of the magnet. It seems to be established that a profound impression had been produced upon the human mind by the natural magnet and its physical properties, the existence of two poles, endowed with opposite properties, and a remote action without direct contact. Even in ancient times it had been observed, or assumed, that the magnet possessed a curative power, and it had been employed as a remedy. This belief still subsisted in the middle ages.[†] In a work by Cardan, dated 1584,[‡] there is an account of an experiment in anæsthesia, produced by the magnet. It was then customary to magnetize rings which were worn round the neck or on the arm, in order to cure nervous diseases. The idea gradually dawned that there are magnetic properties in the human body. The first trace of this belief appears in the works of Paracelsus. This remarkable thinker maintained that

---

[*] Many authors have written the history of animal magnetism : Dubois, Dechambre, Bersot, Figuier, etc. The only study of the subject entitled to be called *critical* is that of Paul Richer, in the *Nouvelle Revue*, August 1, 1882.

[†] Richet, *Bulletin de la Société de Biologie*, May 30, 1884.

[‡] Cardan's Works, book vii., on Precious Stones.

the human body was endowed with a double magnetism; that one portion attracted to itself the planets, and was nourished by them, whence came wisdom, thought, and the senses; that the other portion attracted to itself the elements and disintegrated them, whence came flesh and blood; that the attractive and hidden virtue of man resembles that of amber and of the magnet; that by this virtue the magnetic virtue of healthy persons attracts the enfeebled magnetism of those who are sick.* After Paracelsus, many learned men of the sixteenth and seventeenth centuries—Glocenius, Burgrave, Helinotius, Robert Fludd, Kircher, and Maxwell—believed that in the magnet they could recognize the properties of that universal principle by which minds addicted to generalisation thought that all natural phenomena might be explained. These men wrote voluminous books, filled with sterile discussions, with unproved assertions, and with contemptible arguments.

Mesmer drew largely from these sources; it cannot be disputed that he had read some of these many books, devoted by early authors to the study of magnetism, although such study was forbidden. Where he showed his originality was in taking hold of the so-called universal principle of the world, and in applying it to the sick by means of contact and of passes. His predecessors do not appear to have been addicted to such practices; they believed that in order to infuse the vital spirit, it was enough to make use of talismans and of magic boxes.

Anthony Mesmer was born in Germany, in 1734. He was received as doctor of medicine by the Faculty in

* See Sprengel, *Histoire de la Médecine*, vol. iii. pp. 230 *et seq.*; and Figuier, *Histoire du Merveilleux*, vol. iii. chap. v.

Vienna, and took for the subject of his thesis, *The Influence of the Planets in the Cure of Diseases* (1766). He undertook to prove that the sun, moon, and heavenly bodies act upon living beings by means of a subtle fluid, which he called *animal magnetism*, in order to point out the properties which it has in common with the magnet. After the publication of this whimsical and mystical work, Mesmer made acquaintance with the Jesuit Father Hell, professor of astronomy, who in 1774 settled in Vienna, and cured the sick by means of magnetic steel tractors. Mesmer discovered some analogy between Hell's experiments and his own astronomical theories, and tried what effect the magnet would produce in the treatment of diseases.

An account of his cures filled the Vienna newspapers. Several people of importance gave evidence that they had been cured, among whom was Osterwald, director of the Munich Academy of Science, who had been affected by paralysis; and Bauer, a professor of mathematics, who had suffered from an obstinate attack of ophthalmia. On the other hand, the learned bodies of his native country did not accept his experiments, and the letters which he wrote to most of the academies of Europe remained unanswered. He soon abandoned the use of the magnet and of Hell's instruments, and restricted himself to passes with the hand, declaring animal magnetism to be distinct from the magnet.

Obliged to quit Vienna, in consequence of some adventure not clearly explained, Mesmer came to Paris. He first established himself in a humble quarter of the town, Place Vendôme, and began to expound his theory of the magnetic fluid. In 1779 he published a paper direction, with their necks towards the circumference. All these objects were immersed in water, but this condition was not absolutely necessary, and the *baquet* might be dry. The lid was pierced with a certain number of holes, whence there issued jointed and movable iron branches, which were to be held by the patients. Absolute silence was maintained. The patients were ranged in several rows round the *baquet*, connected with each other by cords passed round their bodies, and by a second chain, formed by joining hands. As they waited a melodious air was heard, proceeding from a pianoforte, or harmonicon, placed in the adjoining room, and to this the human voice was sometimes added. Then, influenced by the magnetic effluvia issuing from the *baquet*, curious phenomena were produced. These are well described by an eye-witness named Bailly:

"Some patients remain calm, and experience nothing; others cough, spit, feel slight pain, a local or general heat, and fall into sweats; others are agitated and tormented by convulsions. These convulsions are remarkable for their number, duration, and force, and have been known to persist for more than three hours. They are characterized by involuntary, jerking movements in all the limbs, and in the whole body, by contraction of the throat, by twitchings in the hypochondriac and epigastric regions, by dimness and rolling of the eyes, by piercing cries, tears, hiccough, and immoderate laughter. They are preceded or followed by a state of languor or dreaminess, by a species of depression, and even by stupor.

"The slightest sudden noise causes the patient to start, and it has been observed that he is affected by a change of time or tune in the airs performed on the pianoforte; that his agitation is increased by a more lively movement, and that his convulsions then become more violent. Patients are seen to be absorbed in the search for one another, rushing together, smiling, talking affectionately, and endeavouring to modify their crises. They are all so submissive to the magnetizer that even when they appear to be in a stupor, his voice, a glance, or sign will rouse them from it. It is impossible not to admit, from all these results, that some great force acts upon and masters the patients, and that this force appears to reside in the magnetizer. This convulsive state is termed the *crisis*. It has been observed that many women and few men are subject to such crises; that they are only established after the lapse of two or three hours, and that when one is established, others soon and successively begin.

"When the agitation exceeds certain limits, the patients are transported into a padded room; the women's corsets are unlaced, and they may then strike their heads against the padded walls without doing themselves any injury."

Mesmer, wearing a coat of lilac silk, walked up and down amid this palpitating crowd, together with Deslon and his associates, whom he chose for their youth and comeliness. Mesmer carried a long iron wand, with which he touched the bodies of the patients, and especially those parts which were diseased; often, laying aside the wand, he magnetized them with his eyes, fixing his gaze on theirs, or applying his hands to the hypochondriac region and to the lower part of the abdomen. This application was often continued for hours, and at other times the master made use of *passes*. He began by placing

himself *en rapport* with his subject. Seated opposite
to him, foot against foot, knee against knee, he laid his
fingers on the hypochondriac region, and moved them to
and fro, lightly touching the ribs. Magnetization with
strong currents was substituted for these manipulations
when more energetic results were to be produced. " The
master, erecting his fingers in a pyramid, passed his
hands all over the patient's body, beginning with the
head, and going down over the shoulders to the feet.
He then returned again, to the head, both back and
front, to the belly and the back; he renewed the process again and again, until the magnetized person was
saturated with the healing fluid, and was transported
with pain or pleasure, both sensations being equally
salutary." *   Young women were so much gratified by
the crisis, that they begged to be thrown into it anew;
they followed Mesmer through the hall, and confessed
that it was impossible not to be warmly attached to
the magnetizer's person.

It must have been curious to witness such scenes.
So far as we are now able to judge, Mesmer excited in
his patients nervous crises in which we may trace the
principal signs of the severe hysteric attacks which may
be observed daily.†   Silence, darkness, and the emotional
expectation of some extraordinary phenomenon, when
several persons are collected in one place, are conditions
known to encourage convulsive crises in predisposed
subjects. It must be remembered that women were in
the majority, that the first crisis which occurred was

---

* Louis Figuier, *Histoire du Merveilleux*, vol. ii. p. 20. Paris, 1860.

† See Bourreville and Regnard, *Iconographie photographique de la Salpêtrière*; Paul Richer, *Études cliniques sur l'Hystero-epilepsia*.

contagious, and we shall fully understand the hysterical character of these manifestations.

We must again draw attention to some of the characteristics of these convulsive crises. The movements of all the limbs and of the whole body, the contraction of the throat, the twitchings of the hypochondriac and of the epigastric regions, are manifest signs of hysteria, and may be referred to the nervous antecedents of the elegant and frivolous crowd which was the subject of Mesmer's experiments. There is, however, still considerable uncertainty as to the nature of many of the phenomena which took place round the *baquet*.

The desire to submit to Mesmer's treatment soon became more general. The house in Place Vendôme became too small, and Mesmer purchased the Hôtel Bullion, in which he established four *baquets*, one of them for the gratuitous use of the poor. Since the latter did not suffice, Mesmer undertook to magnetize a tree at the end of Rue Bondy, and thousands of sick people might be seen attaching themselves to it with cords, in hopes of a cure.

But this rage for Mesmer's treatment could not last long, and difficulties of all kinds assailed him. On his arrival in Paris, he had requested the Academy of Science, and subsequently the Royal Society of Medicine, to institute an inquiry into his experiments; they were unable to agree as to the conditions of this inquiry, and the meeting dissolved in anger. Deslon, a professor of the Faculty of Medicine, asked his colleagues to summon a general meeting to examine his observations and Mesmer's propositions. This meeting, incited by M. de Vauzesmes, was extremely hostile to him. He was condemned without any examination of the facts, and, moreover, was threatened with the penalty of having his name removed from the list of licensed physicians unless he amended his ways. In consequence of this, Mesmer left France, although the government offered him a life-pension of 20,000 francs if he would remain.

Mesmer's absence was short. He was soon recalled by his disciples, who were aware of their master's avarice, and opened a subscription of 10,000 louis, in order to induce him to give a course of lectures in which he was to reveal his discoveries. This course was, however, the point of departure for dissensions between the master and his disciples. Since the latter had bought his secret, they thought themselves entitled to publish it in lectures to the public. Mesmer claimed the monopoly of his discovery. Moreover, in spite of his promises, he had never made a complete explanation, doubtless because he had nothing to tell. He had nothing definite to add to the twenty-seven propositions published in 1779. Several of Mesmer's disciples, who had paid a high price for his secret, accused him of having enunciated a theory which was merely a collection of obscure principles, and in fact they were justified in this assertion. One of Deslon's hearers said: "Those who know the secret are more doubtful than those who are ignorant of it." It was a period of disputes, dissensions, epigrams, invectives, vaudevilles, and songs.

Finally the government intervened, and in 1784 a commission was nominated to inquire into magnetism. This commission consisted of members taken from the Faculty of Medicine, and from the Academy of Sciences. Bailly, the celebrated astronomer, was chosen as its reporter, and it included other illustrious men, such as Franklin and Lavoisier. Another commission, composed of members of the Royal Society of Medicine, was charged to make a distinct report on the same subject; Laurent de Jussieu was included in this second commission.

We find it interesting at the present day to read the reports of these commissions, since they contain a disquisition on an obscure matter, of which time has revealed part of the secret. The line of conduct pursued by the commissioners in their inquiry was irreproachable. The question concerned the existence of a magnetic fluid of the nature which Mesmer and Deslon claimed to have discovered. Deslon proposed to prove the existence of the fluid by the observation of the cures which he effected. But the commissioners rightly considered that this method was too doubtful; they decided to observe in the first instance " the instantaneous effects of the fluid on the animal body, while depriving these effects of all the illusions which might be allied with them, and ascertaining that they could be due to no other cause than animal magnetism."

The immediate effects of magnetism, as they occurred at this period, were *crises*, and these were the special object of research. Some really magnetic effects might be combined with them, but Mesmer and his disciples only ascribed curative virtue to the manifestation of these convulsive movements. Deslon asserted that it was only by means of these crises, which were produced and directed by the will of the magnetizer, that he was able to assist or excite the efforts of nature, and thus effect a cure. We are now aware that these crises are real phenomena, of which the cause is generally admitted to be hysterical neurosis. Moreover, a considerable number of facts demonstrate that, under the influence of such crises, certain forms of paralysis, which have persisted for months, and even for years, may suddenly disappear. There was, therefore, a certain truth in the curative virtue of these convulsive phenonema.

The commissioners placed themselves under treatment once a week, and experienced nothing, except from time to time, after the *séance* had been protracted for several hours, a slight nervous irritability or pain in the hollow of the abdomen, to which Deslon applied his hand. We can understand this negative experience since we are aware that such crises, as well as magnetism, can only be produced in a favourable soil. In the case of susceptible subjects, the commissioners observed an extreme difference between those who were treated in public and in private, and this can be still more readily explained by the well-known contagious effect of example in all hysterical manifestations. The commissioners were particularly struck with the fact that the crises did not occur unless the subjects were aware that they were being magnetized. For instance, in the experiments performed by Jumelin, they observed the following fact. A woman who appeared to be a very sensitive subject, was sensible of heat as soon as Jumelin's hand approached her body. Her eyes were bandaged, she was informed that she was being magnetized, and she experienced the same sensation, but when she was magnetized without being informed of it, she experienced nothing. Several other patients were likewise strongly affected when no operation was taking place, and experienced nothing when the operation was going on. But the most curious experience of this kind was made in Deslon's presence, much to his confusion. According to the theory, when a tree was magnetized, every person who approached it was affected by its influence. The experiment was made at Passy when Franklin was present. Deslon magnetized one tree in an orchard, and a boy of twelve years old, very sensitive to magnetism, was brought towards it with his eyes bandaged. At the first, second, and third tree, he turned giddy; at the fourth, when he was still at a distance of twenty-four feet from the magnetized tree, the crisis occurred, his limbs became rigid, and it was necessary to carry him to an adjoining grass-plat before Deslon could recall him to consciousness. All that these experiments show is that the preconceived idea may produce the same magnetic effects as purely physical means. This truth is well known to the performers of experiments. It is now an established fact that a subject may be thrown into a magnetic sleep, simply by assuring him that this will occur, and by the same process he may even be magnetized from a distance, if it is asserted that he will fall into somnambulism on a given day and hour, in any place which has been selected.

The commissioners, ignorant of all these phenomena, which are now thoroughly established, thought that all which they had observed might be explained by three chief causes—imitation, imagination, and contact. This is the conclusion of their report:—

"The commissioners have ascertained that the animal magnetic fluid is not perceptible by any of the senses; that it has no action, either on themselves or on the patients subjected to it. They are convinced that pressure and contact effect changes which are rarely favourable to the animal system, and which injuriously affect the imagination. Finally, they have demonstrated by decisive experiments that imagination apart from magnetism, produces convulsions, and that magnetism without imagination produces nothing. They have come to the unanimous conclusion with respect to the existence and utility of magnetism, that there is nothing to prove the existence of the animal magnetic fluid; that this fluid, since it is non-existent, has no beneficial effect; that the violent effects observed in patients under public treatment are due to contact, to the excitement of the imagination, and to the mechanical imitation which involuntarily impels us to repeat that which strikes our senses. At the same time, they are compelled to add, since it is an important observation, that the contact and repeated excitement of the imagination which produce the crises may become hurtful; that the spectacle of these crises is likewise dangerous, on account of the imitative faculty which is a law of nature; and consequently. that all treatment in public in which magnetism is employed must in the end be productive of evil results.

"(Signed)   B. FRANKLIN, MAJAULT, LE ROY, SALLIN,
          BAILLY, D'ARCET, DE BORY, GUILLOTIN,
          LAVOISIER.

"Paris, August 11, 1784."

The commissioners therefore merely regarded magnetism as an effect of the imagination. Deslon appears to have come to the same conclusion, since he says, not unreasonably, "If the medicine of the imagination is the most efficient, why should we not make use of it ?"   In

our day this would appear to be an insufficient explanation. We might as well say that hysteria is due to the
imagination.

At the same time, the commissioners presented a
secret report which expressed their final estimate of
magnetism. It is the object of this curious document
to point out the dangers of magnetism with respect to
morality. We think it well to reproduce it *in extenso*.

"The commissioners entrusted by the king with the
examination of animal magnetism have drawn up a
report to be presented to his Majesty which ought
perhaps to be published. It seemed prudent to suppress
an observation not adapted for general publication, but
they did not conceal it from the king's minister. This
minister has charged them to draw up a note designed
only for the eyes of the king.

"This important observation concerns morality. The
commissioners have ascertained that the chief causes of
the effects ascribed to animal magnetism are contact,
imagination, and imitation. They have observed that
the crisis occurs more frequently in women than in
men. The first cause of this fact consists in the different organizations of the two sexes. Women have, as
a rule, more mobile nerves; their imagination is more
lively and more easily excited; it is readily impressed
and aroused. This great mobility of the nerves, since
it gives a more exquisite delicacy to the senses, renders
them more susceptible to the impressions of touch. In
touching any given part, it may be said that they are
touched all over the body, and the mobility of their
nerves also inclines them more readily to imitation. It
has been observed that women are like musical strings

stretched in perfect unison; when one is moved, all the others are instantly affected. Thus the commissioners have repeatedly observed that when the crisis occurs in one woman, it occurs almost at once in others also.

"This organization explains why the crises in women are more frequent, more violent, and of longer duration than in men; it is nearly always due to their sensitive nerves. Some crises are due to a hidden, but natural cause, to an emotional cause to which women are more or less susceptible, and which, by a remote influence, accumulates these emotions and raises them to their highest pitch, thus producing a convulsive state which may be confounded with the ordinary crises. This is due to the empire which nature has caused one sex to exert over the other, so as to arouse feelings of attachment and emotion. Women are always magnetized by men; the established relations are doubtless those of a patient to the physician, but this physician is a man, and whatever the illness may be, it does not deprive us of our sex, it does not entirely withdraw us from the power of the other sex; illness may weaken impressions without destroying them. Moreover, most of the women who present themselves to be magnetized are not really ill; many come out of idleness, or for amusement; others, if not perfectly well, retain their freshness and their force, their senses are unimpaired and they have all the sensitiveness of youth; their charms are such as to affect the physician, and their health is such as to make them liable to be affected by him, so that the danger is reciprocal. The long-continued proximity, the necessary contact, the communication of individual heat, the interchange of looks, are ways and means by which it is well known that nature ever effects the communication of the sensations and the affections.

"The magnetizer generally keeps the patient's knees enclosed within his own, and consequently the knees and all the lower parts of the body are in close contact. The hand is applied to the hypochondriac region, and sometimes to that of the ovarium, so that the touch is exerted at once on many parts, and these the most sensitive parts of the body.

"The experimenter, after applying his left hand in this manner, passes his right hand behind the woman's body, and they incline towards each other so as to favour this twofold contact. This causes the closest proximity; the two faces almost touch, the breath is intermingled, all physical impressions are felt in common, and the reciprocal attraction of the sexes must consequently be excited in all its force. It is not surprising that the senses are inflamed. The action of the imagination at the same time produces a certain disorder throughout the machine; it obscures the judgment, distracts the attention; the women in question are unable to take account of their sensations, and are not aware of their condition.

"The medical members of the commission were present to watch the treatment, and carefully observed what passed. When this kind of crisis is approaching, the countenance becomes gradually inflamed, the eye brightens, and this is the sign of natural desire. The woman droops her head, lifts her hand to her forehead and eyes in order to cover them; her habitual modesty is unconsciously aroused, and inspires the desire of concealment. The crisis continues, however, and the eye is obscured, an unequivocal sign of the complete disorder of the senses.

This disorder may be wholly unperceived by the woman who experiences it, but it cannot escape the observant eye of the physician. As soon as this sign has been displayed, the eyelids become moist, the respiration is short and interrupted, the chest heaves rapidly, convulsions set in, and either the limbs or the whole body is agitated by sudden movements. In lively and sensitive women this last stage, which terminates the sweetest emotion, is often a convulsion; to this condition there succeed languor, prostration, and a sort of slumber of the senses, which is a repose necessary after strong agitation.

"This convulsive state, however extraordinary it may appear to the observers, is shown to have nothing painful or contrary to nature in it, from the fact that, as soon as it is over, it leaves no unpleasant traces in its subjects. There is nothing disagreeable in the recollection, but, on the contrary, the subjects feel the better for it, and have no repugnance to enter anew into the same state. Since the emotions they experience are the germs of the affections and inclinations, we can understand why the magnetizer inspires such attachment, an attachment likely to be stronger and more marked in women than in men, so long as men are entrusted with the task of magnetism. Undoubtedly many women have not experienced these effects, and others have not understood the cause of the effects they experienced; the more modest they are, the less they would be likely to suspect it. But it is said that several have perceived the truth, and have withdrawn from the magnetic treatment, and those who have not perceived it ought to be deterred from its pursuit.

"The magnetic treatment must necessarily be danger-ous to morality. While proposing to cure diseases which require prolonged treatment, pleasing and precious emotions are excited, emotions to which we look back with regret and seek to revive, since they possess a natural charm for us, and contribute to our physical happiness. But morally they must be condemned, and they are the more dangerous as it becomes more easy for them to become habitual. A condition into which a woman enters in public, amid other women who apparently have the same experience, does not seem to offer any danger ; she continues in it, she returns to it, and discovers her peril when it is too late. Strong women flee from this danger when they find themselves exposed to it; the morals and health of the weak may be impaired.

"Of this danger M. Deslon is aware. On the 9th of last May, at a meeting held at M. Deslon's own house, the lieutenant of police asked him several questions on this point in the presence of the commissioners. M. Lenoir said to him, 'In my capacity as lieutenant-general of police, I wish to know whether, when a woman is mag-netized and passing through the crisis, it would not be easy to outrage her.' M. Deslon replied in the affirmative, and it is only just to this physician to state that he has always maintained that he and his colleagues, pledged by their position to act with probity, were alone entitled and privileged to practise magnetism. It must be added that although his house contains a private room origin-ally intended for these crises, he does not allow it to be used. The danger exists, however, notwithstanding this observance of decency, since the physician can, if he will, take advantage of his patient. Such occasions may occur daily and at any moment; he is sometimes exposed to the danger for two or three hours at a time, and no one can rely on being always master of his will. Even if we ascribe to him superhuman virtue, since he is exposed to emotions which awaken such desires, the imperious law of nature will affect his patient, and he is responsible, not merely for his own wrong-doing, but for that he may have excited in another.

"There is another mode of producing convulsions, a mode of which the commissioners have obtained no direct and positive proof, but which they cannot but suspect; namely, a simulated crisis, which is a signal for, or produces many others, out of imitation. This expedient is, at any rate, needed to hasten or maintain the crises which are an advantage to magnetism, since without them it could not be carried on.

"There are no real cures, and the treatment is tedious and unprofitable. There are patients who have been under treatment for eighteen months or two years without deriving any benefit from it; at length their patience is exhausted, and they cease to come. The crises serve as a spectacle; they are an occupation and interest, and, moreover, they are to the unobservant the result of magnetism, a proof of the existence of that agent, although they are really due to the power of the imagination.

"When the commissioners began their report, they only stated the result of their examination of the magnetism practised by M. Deslon, to which the order of the king had restricted them, but it is evident that their experiments, observations, and opinions apply to magnetism in general. M. Mesmer will certainly declare that the commissioners have not examined his method, proceedings, and the effects they have produced. The commissioners are undoubtedly too cautious to pronounce on that which they have not examined, and with which they are not acquainted, yet they must observe that M. Deslon's principles are those of the twenty-seven propositions printed by M. Mesmer in 1779.

"If M. Mesmer has enlarged his theory, it thereby becomes more absurd: the heavenly influences are only a chimæra, of which the fallacy has long been recognized. The whole theory may be condemned beforehand, since it is based upon magnetism; and it has no reality, since the animal magnetic fluid has no existence. Like magnetism, this brilliant theory exists only in the imagination. M. Deslon's mode of magnetizing is the same as that of M. Mesmer, of whom he is the disciple. When we place them together, we see that they have treated the same patients, and, consequently, have pursued the same process: the method now in use by M. Deslon is that of M. Mesmer.

"The results also correspond; the crises are as violent and frequent, and the same symptoms are displayed under the treatment of M. Deslon and of M. Mesmer. Although the latter may ascribe an obscure and inappreciable difference to his method, the principles, practice, and results are the same. Even if there were any real difference, no benefit from such treatment can be inferred, after the details given in our report and in this note, intended for the king.

"Public report declares that M. Mesmer's cures are not more numerous than those of M. Deslon. There is nothing to prevent the convulsions in this case also from becoming habitual, from producing an epidemic, and from being transmitted to future generations : such practices and assemblies may also have an injurious effect upon morality.

"The commissioners' experiments, showing that all these results are due to contact, to imagination and imitation, while explaining the effects produced by M. Deslon, equally explain those of M. Mesmer. It may, therefore, reasonably be concluded that, whatever be the mystery of M. Mesmer's magnetism, it has no more real existence than that of M. Deslon, and that the proceedings of the one are not more useful nor less dangerous than those of the other.

"(Signed)   FRANKLIN, BORY, LAVOISIER, BAILLY, MAJAULT, SALLIN, D'ARCET, GUILLOTIN, LE ROY.

"Paris, August 11, 1784."

The Royal Society of Medicine presented their report five days later, and came to the same conclusions. But one member of the commission, Laurent de Jussieu, dissented from his colleagues, and, with scientific courage, published a separate report, containing his convictions on the subject.

De Jussieu had performed some experiments which could not, as he thought, be explained by the imagination. These facts demonstrated, in his opinion, that man produced a sensible action upon his fellow by friction, by contact, and, more rarely, by simple proximity. This action, ascribed to an universal fluid not yet demonstrated, was, he said, certainly due to animal heat, which he elsewhere terms animalized electric fluid. With respect to the theory of animal magnetism, he did not reject it as absolutely as Bailly, who said, "Everything is done by

the imagination; magnetism has nothing to do with it."
He was content with saying, much more wisely, that the
theory of magnetism could only be accepted when it was
developed and supported by substantial proofs. In short,
as Dechambre remarks, the idea pervades this report
that Mesmer is on the track of a fruitful truth. This
presentiment of the illustrious naturalist was soon to be
confirmed; and, moreover, it is worth while to consider
some of the assertions in de Jussieu's paper, since they
contain an element of truth.

The efficacy of the action of contact and friction is
proved by the existence in certain subjects of hypnogenic
zones, of which the slightest stimulation produces somnambulism. M. Charcot has shown that the irritation of
hysterogenic zones produces convulsions, and these zones
are generally seated in the hypochondriac, or in the
ovarian regions, on which Mesmer preferred to exercise
his manipulations.

After Bailly's report, Mesmer left France, and returned
to Germany. His part was played out, and we shall not
recur to it. His friends have represented him as a man
desirous of fame, but at the same time full of love for
suffering humanity. Public opinion, more severe in its
judgment, regards him as the type of the scientific
charlatan.

Up to this time, animal magnetism had not been discovered; it probably had something to do with most of
the mesmeric phenomena, with the *baquet*, etc.; but it was
not recognized amid the nervous crises excited by Mesmer.
It is to one of his disciples, to the Marquis Armand Jacques
Marc Chastenet de Puységur, that the discovery must
be ascribed of animal magnetism, or of artificial som-

nambulism, which ought, therefore, to bear the name of Puységurian somnambulism.*

In May, 1784, M. de Puységur, living in retirement on his estate of Buzancy, near Soissons, employed his leisure in magnetizing peasants, after the manner of his master, and on one occasion he chanced to observe the production of an entirely new phenomenon.† A young peasant named Victor, twenty-three years of age, who had been suffering for four days from inflammation of the lungs, was thrown by magnetism into a peaceful sleep, unaccompanied by convulsions or suffering. He spoke aloud, and was busied about his private affairs. It was easy to change the direction of his thoughts, to inspire him with cheerful sentiments, and he then became happy, and imagined that he was firing at a mark or dancing at a village *fête*. In his waking state he was simple and foolish, but during the crisis his intelligence was remarkable; there was no need of speaking to him, since he could understand and reply to the thoughts of those present. He himself indicated the treatment necessary in his illness, and he was soon cured.

This is a brief account of the peasant Victor's case. The news of his cure was rapidly spread abroad, and from all sides there was a concourse of sick people demanding relief. The phenomenon was repeated, to de

---

* The following are the works of Puységur:—*Mémoires pour servir à l'Histoire du Magnétisme animal*, 1784; *Suite aux Mémoires*, 1805; *Du Magnétisme animal*, etc., 1807; *Recherches, Expériences et Observations physiologiques sur l'homme, dans l'état de somnambulisme naturel, et dans le somnambulisme provoqué par l'acte magnétique*, 1811; etc.

† Puységur asserts that Mesmer must have been acquainted with somnambulism, but that he did not choose to mention his discovery to his disciples.

Puységur's great joy, and he wrote: "My head is turned with joy, now that I see what good I am doing." Since he was unable to minister to the continually increasing number of patients, the marquis pursued Mesmer's plan of magnetizing an elm which grew on the village green at Buzancy. The patients were seated on stone benches round this tree, with cords connecting its branches with the affected parts of their bodies, and they formed a chain by linking their thumbs together. Meanwhile de Puységur chose from among his patients several subjects who, through contact with his hands or on the presentation of a metallic tractor, fell into the ordinary crisis, and this soon passed into a sleep in which all physical faculties appeared to be suspended, while the mental faculties were enlarged.

Cloquet, an eye-witness,* has given us some valuable information on the subject. He says that the patient's eyes were closed, and there was no sense of hearing, unless it was awakened by the master's voice. Care was taken not to touch the patient during his crisis, nor even the chair on which he was seated, as this would produce suffering and convulsions, which could only be subdued by the master. To rouse them from the trance, the master touched the patient's eyes, or said, "Go and embrace the tree." Then they arose, still asleep, went straight to the tree, and soon afterwards opened their eyes. As soon as they returned to a normal condition, the patients retained no recollection of what had occurred during the three or four hours' crisis.

But it was the cure of diseases at which de Puységur

* *Détails des cures opérées à Buzancy, près Suissons par le magnétisme animal.* Soissons: 1784.

aimed : therapeutics were his object, as it had been that of Mesmer. He observed, or thought that he observed, that during the crisis, the patients possessed a supernatural power which entitled them to be called physicians; it was, in fact, enough for them to touch through his clothes the sick person presented to them, in order to feel the part affected, and to indicate fitting remedies. Since they were solely occupied with this question, de Puységur and the other magnetizers who followed his example in Lyons, Bordeaux, Bayonne, Marseilles, etc., did not study the natural history of this artificial sleep. De Puységur, like Mesmer, was a healer. But in the case of de Puységur's treatment we agree with Dechambre that if his faith was robust, so likewise was his honesty. There was no public exhibition, nothing was done to strike the imagination; there was no selection of subjects from among silly or melancholic women. His patients of both sexes were of the peasant class, and were often suffering from severe and obstinate diseases. De Puységur's honesty and disinterestedness contrast well with Mesmer's avarice.

As far as de Puységur's theoretic views are concerned, they are slight modifications of those of Mesmer. As little versed in physical science as his master, he always maintains the existence of an universal fluid, of which he recognizes the *electric* nature; this fluid saturates all bodies, and especially the human body, which has a perfect electric organization, and is an animated electric machine. Man can display this electric fluid at pleasure, and diffuse it externally by his movements, in order to produce somnambulism. It is curious that de Puységur should have strongly condemned the use of magnets in the treatment of disease, and of all electricity foreign to our organism. This dogma has been falsified, and, as we are aware, electro-therapeutics has come into use.

In this way de Puységur modified the tradition he had received from Mesmer, and simple contact or spoken orders were substituted for the use of the *baquet*. There were no more violent crises, accompanied by cries, sobs, and the contortions of an attack of hysteria ; instead of these, there was a calm, peaceful, healthy, and composed slumber. This was not a transformed phase of magnetism, but the actual discovery of this state, of which the honour is due to de Puységur.

It is easy to disentangle the portion of truth which exists in the. descriptions of the magnetic sleep left by de Puységur. He has carefully observed the obedience of the magnetized subjects to the magnetizer's orders, who directs their thoughts and acts at his pleasure. We shall presently study this symptom under the name of *suggestion*. He has also observed the patient's unconsciousness, and that he retains no recollection of what has occurred during sleep. We shall see that this unconsciousness is a frequent and almost constant phenomenon during profound hypnotism. Finally, the descriptions show the singular affinity which seems to exist between the magnetizer and his subject; a phenomenon which is shown in some curious ways : the magnetizer alone must touch the sleeping subject, for fear of producing suffering and even convulsions. All this is accurate, established by science, and now admitted by every one. But it is not yet admitted that the subject is able to divine the thoughts of the magnetizer without any material communication, nor that the patient is acquainted with the nature of his disease, and can indicate effectual

remedies and foresee future events. De Puységur tried
to give this faculty an air of probability by naming it
pressentation.

Mesmer's theory had been condemned by the judgment of scientific bodies, and this judgment was not
reversed by de Puységur's experiments, in which there
was too much of the supernatural. Professional magnetizers adopted his experiments as the theme of their
discourses. We can also understand the favour with
which his assertion of the clairvoyance of somnambulists was received, since this was a new form of the
gift of divination which had always obtained credence.
Numerous magnetic societies were formed in different
parts of France, especially, as Thouret states, in those
towns which possessed no university, and which were
therefore less under control. The Harmonic Society,
however, founded at Strasburg, consisted of more than
one hundred and fifty members.

We must mention in passing Pététin's experiments
in catalepsy, since he had the good fortune to be the
first, or one of the first, to observe the phenomena of the
transposition of the senses. Pététin was a Lyons
physician, President of the Medical Society in that city,
and opposed to the new theories of magnetism. He
observed and exhibited to his colleagues a cataleptic
woman who saw, heard, felt, smelled, and tasted by
means of the epigastric region and of the finger-tips.
This occurred in 1787. After Pététin's death a paper
by him appeared, containing seven observations of the
same kind. He ascribed these strange phenomena to
the accumulation of the animal electric fluid in certain
parts of the body. The magnetizers seized upon this fact,

and we shall see that for some time to come the question of the transposition of the senses was predominant.

Up to the year 1820, we find no work to quote, except that of the naturalist Deleuze, on the history of magnetism. His book is entitled *Histoire critique du Magnétisme animal* (1813); it is a crude work, which has been estimated much above its real value, and while it displays the honesty and sincerity of its author, it adds nothing to the sum of our knowledge on the subject.

Deleuze, like his predecessors, was chiefly concerned with the curative virtues of magnetism; and in order to prove their reality, he found no better expedient than to advise the incredulous to make use of it in various diseases. He said, moreover, that faith was essential to success, thus dispensing with any legitimate demonstration. Magnetism was held to be applicable to all diseases, and constituted, as in the days of de Puységur and of Mesmer, an universal panacea. At about the same period, in 1813, a thaumaturgist named Faria, who came from the Indies, gave public representations, for money, of the wonders which could be effected by means of magnetism. The process by which he induced sleep was curious. He seated the subject in an armchair, with closed eyes, and then cried out in a loud and imperious voice, "Go to sleep!" After a slight movement, the subject sometimes fell into a condition which Faria termed a lucid slumber. This charlatan had rightly observed that the cause of somnambulism rests in the subject himself. He truly said that sleep might be induced at the will of the subject, or when such will was absent, or even when it was exerted in the contrary sense.

CHAPTER II.

In 1820 it might have been supposed that animal
magnetism was about to enter upon a scientific era. Dr.
Bertrand, a former pupil of the Polytechnic School, had
just brought the subject before the public in a course
of lectures. General Noizet, about the same time, drew
up a paper for the Royal Academy of Berlin on somnambulism and animal magnetism. Experiments were performed in the hospitals, directed at the Hôtel-Dieu by
Du Potet, pupil of Husson, and at the Salpêtrière by
Georget and Rostan. The experiments made on hysterical
patients were not such as to modify the scepticism of
the scientific world, and it was thought probable that
the experimenters had been deceived by their patients.
Indeed Pétronille, one of Georget's well-known somnambulists, afterwards confessed that she had imposed on
the observers. But Richer justly observes that such
confidences are the common boasts of hysterical patients,
and that those who believe them incur the same reproach
of credulity as their opponents are charged with.

The general council of the hospitals put an end to
these operations, on the ground that the patients should

not be subjected to such experiment, but on all sides the need of some definite proof was felt.

In 1825 Foissac induced the Academy of Medicine, which had succeeded to the Royal Society of Medicine, to take part in the controversy. He drew up a paper, in which he undertook to show that simple contact enabled his somnambulists to diagnose their diseases, with an intuition *worthy of the genius of Hippocrates.* Although such language did not seem to be adapted to convince the Academy, its members nominated a commission charged to decide whether it was expedient to undertake a fresh examination into the question of animal magnetism. The report presented by Husson was in favour of such an examination, and the Academy, by a majority of thirty-five votes against twenty-five, nominated a commission of inquiry, consisting of Bourdois, Double, Fouquier, Itard, Guéneau de Mussy, Guersant, Leroux, Magendie, Marc, Thillaye, and Husson. Magendie and Double, finding that the experiments were not very carefully performed, took no part in the labours of the commission. At the end of five years' patient research, in June, 1831, Husson presented a report in which the existence of animal magnetism was affirmed. "The results are negative or insufficient in the majority of cases," the report declares; "in others they are produced by weariness, monotony, or by the imagination. It appears, however, that some results depend solely on magnetism, and cannot be produced without it. These are physiological phenomena, and well established therapeutically." The importance of this work decides us to reproduce its principal conclusions *in extenso.*

"The contact of the thumbs and hands, friction, or

the employment of certain gestures within a short distance of the body, which are called passes, are the means employed to place the patient *en rapport*, or, in other words, to transmit the action of the magnetizer to his subject.

"The time necessary for transmitting and effecting this magnetic action varies from half an hour to one minute.

"When once a person has been thrown into the magnetic sleep, it is not always necessary to have recourse to contact and passes in order to magnetize him afresh. A glance from the magnetizer, or his will alone, may have the same influence.

"The effects produced by magnetism are extremely varied; it agitates some people and calms others; it generally causes a momentary quickening of the respiration and of the circulation; this is followed by fibrillary, convulsive movements like those produced by electric shocks; by a more or less profound torpor; by stupor and somnolence; and, in a few instances, by what magnetizers term somnambulism.

"The perceptions and faculties of individuals who are thrown by magnetism into a state of somnambulism are modified in various ways.

"Some, amid the noise of general conversation, only hear the voice of their magnetizer; many make a direct reply to the questions which he or the persons with whom they are placed *en rapport* address to them; others converse with all those who surround them; in few instances are they aware of what is passing. They are generally completely unconscious of any sudden external noise made close to their ears, such as the striking of copper vessels, the fall of a piece of furniture, etc.

'·The eyes are closed, and the lids yield with difficulty to any effort made with the hand to open them. This operation causes pain, and the pupil of the eye is then seen to be contracted and turned upwards, or sometimes towards the base of the orbit.

"Sometimes the sense of smell is altogether absent, and they may be made to breathe nitric acid or ammonia without being incommoded, or even without their becoming aware of it. But this is not always the case, and some subjects retain the sense of smell.

"Most of the somnambulists whom we have observed were completely insensible. The feet might be tickled, the nostrils and the corner of the eyelid might be touched with a feather, the skin might be pinched until it was discoloured, pins might suddenly be driven to some depth under the nails, and the subjects would betray no sign of pain, nor even a consciousness of the fact. Finally, a somnambulist has been rendered insensible to one of the most painful surgical operations, and neither the countenance, the pulse, nor the respiration betrayed the slightest emotion.

"We have only observed one individual who was thrown into the state of somnambulism when magnetized for the first time. Sometimes somnambulism only occurs after the eighth or tenth *séance*.

"We have constantly observed that natural sleep, which is the repose of the organs of the senses, of the intellectual faculties, and of voluntary movements, precedes and terminates the state of somnambulism.

"The magnetized subjects whom we have observed

under somnambulism retain the faculties of the waking state. The memory even appears to be more retentive and of wider range, since they recollect all that occurred on each previous occasion when they were under somnambulism.

"We have observed two somnambulists who were able, with closed eyes, to distinguish the objects placed before them; who could declare, without touching them; the suit and value of playing cards; who could read words traced with the hand, or some lines from a book opened at random. This phenomenon has even occurred when the fingers are firmly pressed upon the closed eyelids.*

* "On January 12 there was a meeting of the commission at the house of M. Foissac. This physician announced that he should put Paul to sleep; that when he was in this state of somnambulism, a finger would be applied to each closed eyelid, and that in spite of this he would distinguish the colour of cards, he would read the title of a book, or some words or lines indicated at random in the book itself. After the magnetic passes had been made for two minutes, Paul was thrown into sleep. The eyelids were kept constantly closed, in turn by Fouquier, Itard, Marc, and the reporter, and a new pack of cards was presented to him, from which the royal stamp was freshly removed. When these were shuffled together, Paul named them successively without effort: the king of spades, the ace of clubs, the queen of spades, the nine of clubs, the seven, the queen, and the eight of diamonds.

" When the eyelids were kept closed by Ségalas, a volume with which the reporter was provided was presented to him. He read from the title-page, *Histoire de France*, was unable to read the two intermediate lines, and could read only the name of *Anquetil* on the fifth line, where it is preceded by the preposition *par*. The book was then opened at page 88, and he read the first line: '*le nombre de ses . . .*' He missed the word *troupes*, and went on, '*Au moment où on le croyait le plus occupé des plaisirs du carnaval.*' He likewise read the running title *Louis*, but was unable to read the Roman figures which followed it. A paper was presented him on which were written the words *agglutination* and *magnétisme animal*. He spelled the first word, and pronounced the two others. Finally, the report of this *séance* was presented to him; he read the date

3

"In two somnambulists we observed the power of foreseeing the more or less remote or complicated acts of the organism. One of them announced, several days, and even months, in advance, the day, hour, and minute on which an epileptic attack would occur; the other indicated the epoch of his cure. Their previsions were verified with remarkable accuracy. These appear only to apply to the acts and lesions of their own organisms.

"We only observed one somnambulist who indicated the symptoms of the diseases of three persons with whom he was placed *en rapport*, although we inquired into a considerable number of cases.*

with some distinctness, and some of the words which were more legibly written than the rest. In all these experiments the fingers were applied to the whole surface of each eye, by pressing the upper on the lower lid from above in a downward direction, and we observed that there was a constant rotatory movement of the eyeball, as if it were directed towards the object presented to the vision."—*Text of the Report.*

* "M. Marc, a member of the commission, consented to undergo examination by a somnambulist, and Mlle. Céline was requested to consider attentively the state of our colleague's health. She applied her hand to his forehead and to the region of the heart, and at the end of three minutes she said that there was a determination of blood to the head, and that on its left side M. Marc was now suffering from pain; that he was often oppressed, especially after eating; that he was subject to a hacking cough; that the lower part of the chest was congested with blood; that there was obstruction to the passage of food; that there was a contraction in the region of the ensiform appendix; and that in order to effect a cure, M. Marc should be frequently bled, that hemlock plasters should be applied, that he should be rubbed with laudanum on the lower part of the chest, that he should drink lemonade prepared with gum Arabic, that he should eat little and often, and not go out walking immediately after meals.

"We were anxious to hear whether M. Marc's experience agreed with the somnambulist's assertions. He said that he really suffered from oppression after eating, that he was subject to a cough, and had pain on the right side of the head, but that he was not conscious of any uneasiness in the digestive canal.

"We were struck by the analogy between M. Marc's sensations and the

"Some of the magnetized patients experienced no benefit. Others derived more or less relief from the treatment; in one case habitual suffering was suspended, in another strength returned, in a third epileptic attacks were averted for several months, and in a fourth serious paralysis of long standing was completely cured.

"Considered as the agent of physiological phenomena, or as a therapeutic expedient, magnetism must take its place in the scheme of medical science, and consequently it should be practised or superintended by physicians only, which is the rule in northern countries.

"The commission has had no opportunity of verifying the other faculties which are said by magnetizers to be possessed by somnambulists. But the facts collected and now set down, are of sufficient importance to justify the belief that the Academy ought to encourage researches into magnetism, since it is an interesting branch of psychology and of natural history.

"(Signed) BOURDOIS DE LA MOTTE, FOUQUIER, GUÉNEAU DE MUSSY, GUERSANT, ITARD, J. LEROUX, MARC, THILLAYE, HUSSON (reporter)."

Such was the celebrated report, of which the magnetizers made so much that the Academy did not venture to print it.

It must be admitted that the commissioners did not pursue in their researches a rigorously scientific method. Since they were chiefly desirous to prove the existence

assertions of the somnambulist: we noted it carefully, and await a future opportunity of confirming the existence of this singular faculty."—*Text of Report.*

or non-existence of animal magnetism, they applied themselves almost exclusively to the study of extraordinary facts. They thought that if the results of a given experiment exceeded the limits of the possible, animal magnetism would thereby be proved. In this way the question was wrongly stated, since it was possible that magnetism might be at once a natural fact, and a fact which agreed with known physiological laws. The commissioners did not understand this elementary truth. Impelled by curiosity with respect to the marvellous and the supernatural, they directed their attention to those phenomena which were the most disputed and the most open to dispute, such as the transposition of the senses, the power of reading with bandaged eyes or vision by means of the internal organs, by the epigastrium or the occiput, together with the diagnosis of diseases and an acquaintance with their remedies.

It appears that on all these points the conduct of the inquiry was unsatisfactory, and that the commissioners neglected to take any sufficient precautions. Some of the experiments were really futile. The report states that a somnambulist named Petit, whose eyes were so firmly closed that the eyelashes were interlaced, and who was constantly watched by commissioners who "held the light," was able to read what was presented to him, and played several games of piquet with great spirit. It does not appear that any precautions were taken to prevent this individual from reading through his eyelashes. The commissioners were content to watch his eyes, and it did not occur to them that there is nothing more easy than to read with the eyes apparently closed.

At another *séance*, Paul, a young law-student, over whose eyes a commissioner placed his hand, displayed a marvellous clairvoyance; he divined the cards in a pack and could read almost fluently. The reporter observed, however, that the eyeball was constantly rolling, and appeared to be directed towards the object presented to the vision. When we add that the young man read slowly, before a large circle, and that he made mistakes, we shall agree with Ségalas, a member of the Academy, who had himself on one occasion kept the eyes of the subject closed, that it was probably possible to move the eyelids, to catch a glimpse of some of the words, and to guess the rest. At any rate, more careful experiments were needed before admitting that it is possible to see and read with closed eyes. We do not speak of internal vision, of the prevision of crises, and the instinctive knowledge of remedies, since the experiments were all of the same stamp.

Together with these unsatisfactory statements, we find some good descriptions of somnambulism. The commissioners observed that when the subjects were put to sleep they presented "an acceleration of the pulse and of the breathing, fibrillary movements like those produced by electric shocks, stupor, and somnolence. . . . The subject sometimes made a direct reply to the question addressed to him, but in general he was quite unconscious of any sudden noise made at his ear. . . . The eyes were closed, and on raising the eyelid, the pupil was seen to be contracted and turned upwards. . . . The surface of the body was generally insensible to pain; . . . the skin might be pinched until it was discoloured, pins might be driven under the nails without disturbing the subject's impassibility." All this description is excellent: it is unfortunate that the commissioners, who observed the natural phenomenon with such accuracy, were unable to detach it from the phantasmagoria by which it was surrounded.

Finally, the commissioners were mistaken in two points. First, in confounding the question of animal magnetism with the extraordinary and supernatural phenomena described by the magnetizers; secondly, in not bringing to a study of these phenomena, which required the utmost caution, the rigorous care which we have a right to demand from an academical commission.

The Academy, which did not include among its members many partisans of magnetism, was somewhat astonished by Husson's report. It was read in the meetings held on the 21st and 28th of June, 1831. But there was no public debate, nor was the question put to the vote. The report was not even printed, only committed to writing. The Academy shrank from deciding such burning questions.

In 1837 the brooding discussion burst forth, on account of the painless extraction of a tooth during the magnetic sleep, which was related by M. Oudet.

Berna, a young magnetizer, implored the attention of the Academy of Medicine, and a fresh commission was nominated. It consisted of Roux, Bouillaud, Cloquet, Emery, Pelletier, Caventou, Cornac, Oudet, and Dubois, the last-named acting as reporter. The Academy was again drawn in the wrong direction. Berna urged them to examine extraordinary phenomena, such as vision without using the eyes, and the communication of the magnetizer's thoughts to his subject, phenomena which he boasted of producing in two of his somnambulist subjects.

The results of this inquiry, which was conducted with greater care than that of the previous commission, were negative. We give the conclusions of this report, as we have already given those of Husson's report.

"1st *Conclusion.*—Dubois, in terminating his report, states that it appears from all the facts and incidents witnessed by us that, in the first place, no special proof has been given to us as to the existence of a special state, called the state of magnetic somnambulism; that it is only by way of assertion, and not by way of demonstration, that the magnetizer has affirmed at each *séance*, before undertaking any experiments, that his subjects were in a state of somnambulism.

"It is true that, according to the magnetizer's programme, we might be assured that the subject, before he was thrown into a state of somnambulism, was in perfect possession of all his senses, that for this purpose we were to prick him, and that he would then be put to sleep in the presence of the commissioners. But it appeared from our experiments at the *séance* of the 3rd of March, and before any magnetizing process had taken place, that the subject of experiment was as insensible to pin-pricks before the supposed sleep as he was when it had occurred; that his countenance and replies varied little before and after the so-called magnetic sleep. Your commissioners are unable to decide whether this was from inadvertence, from a natural or acquired insensibility to pain, or from an unreasonable desire to attract attention. It is true that we were told on each occasion that the subjects were asleep, but this was purely a matter of assertion.

"If, however, experiments made upon subjects presumed to be in a state of somnambulism should ultimately prove the existence of such a state, the conclusions we are about to draw from their experiments will show whether such proofs have any value or not.

"*2nd Conclusion.*—According to the terms of the programme, the second experiment is intended to establish that the subjects are insensible to pain.

"We must, however, recall the restrictions imposed on your commissioners. The face was not to be subjected to such experiments, nor yet those parts of the body which are usually covered, so that they could only be performed on the hands and the neck. These parts were not to be pinched nor twitched, nor placed in contact with any burning substance, nor exposed to any high temperature; the only thing permitted was to insert the points of needles to the depth of half a line, and at the same time the face was half covered by a bandage which did not allow us to observe the expression of the countenance, when the attempt was made to inflict pain. When we recall all these restrictions, we deduce from them the following facts:—(1) that the sensations of pain we were permitted to excite were extremely slight and of limited extent; (2) that they could only be excited on a small portion of the body, which was perhaps accustomed to receive such impressions; (3) that since these impressions were always of the same kind, they were of the nature of *tattooing*; (4) that the face, and particularly the eyes, in which the expression of pain is most apparent, were concealed from the commissioners; (5) that under these circumstances, impassibility, however absolute and complete, could not be accepted by us as a conclusive proof that the subject in question was devoid of sensibility.

"*3rd Conclusion.*—The magnetizer undertook to prove to the commissioners that, by the mere exercise of the will, he had the power of making his subject either locally or generally sensible to pain, which he terms the *restitution* of sensibility.

"As, however, he had been unable to give us any experimental proof that he had taken away and destroyed this girl's sensibility, this experiment was correlative with the other, and it was consequently impossible to prove such a restitution; moreover, the facts observed by us showed that all the attempts made in this direction had completely failed. You must remember, gentlemen, that the only verification consisted in the somnambulist's assertions. When, for instance, she assured the commissioners that she was unable to move her left leg, this was no proof that the limb was magnetically paralyzed; even in this case her words were not in accordance with her magnetizer's pretensions, so that we only obtain assertions without proof, opposed to other assertions, equally without proof.

"*4th Conclusion.*—What we have just said with reference to the abolition and restitution of sensibility, is applicable in every respect to the so-called abolition and restitution of the power of movement, of which your commissioners did not obtain the slightest proof.

"*5th Conclusion.*—One paragraph of the programme is entitled, 'Obedience to the mental order to cease, in the midst of a conversation, to reply verbally and by signs to a given person.'

"In the *séance* of March 5, the magnetizer attempted to prove to the commissioners that the power of his will went so far as to produce this effect: but it resulted from the facts which occurred during this *séance* that, on the contrary, the somnambulist was still unable to hear when the experimenter no longer wished to prevent her from hearing, and that she appeared to possess the power of hearing when he distinctly desired her to hear nothing. So that, according to the somnambulist's assertions, the faculty of hearing, or of ceasing to hear, was in this instance in absolute revolt against the will of the magnetizer.

"But well-considered facts lead the commissioners to the conclusion that there was neither a revolt nor a submission of the will; only an absolute independence.

"*6th Conclusion.*—Transposition of the sense of sight. —The magnetizer, as you are aware, complied with the commissioners' request in turning from the study of the abolition and restitution of sensibility and the power of movement, in order to consider more important facts; namely, the facts of vision without the aid of the eyes. All the incidents in connection with these facts have been shown to you; they occurred in the *séance* of April 3, 1837.

"Berna undertook to show the commissioners that a woman, influenced by his magnetic manipulations, could decipher words, distinguish playing cards, and follow the hands of a watch, not by means of her eyes, but by her occiput—a fact which would imply either the transposition or the inutility of the organs of sight during the magnetic state. These experiments were made, and, as you are aware, were a complete failure.

"All which the somnambulist knew, all which she was able to infer from what was said in her immediate vicinity, all which she could naturally surmise, she uttered with bandaged eyes; from which we at once concluded that she was not without ingenuity. Thus, when the magnetizer invited one of the commissioners to write a word on a card, and to present it to the woman's occiput, she said that she saw a card, and even the writing on the card. If she was asked how many persons were present, she could, since she had seen them enter, approximately declare their number. If she was asked whether she saw a commissioner sitting near her, engaged in writing with a scratching pen, she raised her head, tried to see under the bandage, and said that this gentleman held something white in his hand. When asked whether she saw the mouth of the same individual, who had left off writing and placed himself behind her, she said that he had something white in his mouth. Hence we concluded that this somnambulist, more experienced and adroit than the former one, was able to make more plausible surmises.

"But with respect to facts really adapted to establish vision by means of the occiput, decisive, absolute, and peremptory facts, they were not only altogether absent, but those which we observed were of a nature to give rise to strange suspicions as to this woman's honesty, as we shall presently observe.

"*7th Conclusion.*—Clairvoyance.—When the magnetizer despaired of proving to the commissioners the transposition of the sense of sight, the nullity and superfluity of the eyes during the magnetic state, he sought to take refuge in the fact of clairvoyance, or of vision through opaque bodies.

"You are acquainted with the experiments made on this subject. The main conclusion deduced from these facts was that a man, placed before a woman in a given attitude, is unable to give her the power of distinguishing the objects presented to her when her eyes are bandaged.

"Here your commissioners were occupied with a more serious reflection. Admitting for a moment an hypothesis which is very convenient for magnetizers, that in many cases somnambulists lose all lucidity, and are as unable as ordinary mortals to see by means of the occiput, of the stomach, or through a bandage, what are we to conclude with respect to the woman who gave minute description of objects quite different from those presented to her? We are at a loss what to think of a somnambulist who described the knave of clubs on a blank card, who transformed the ticket of an academician into a gold watch with a white dial-plate inscribed with black figures, and who, if she had been pressed, would perhaps have gone on to tell us the hour marked by this watch. . . .

"If, gentlemen, you now ask what is the ultimate and general conclusion to be inferred from all these experiments, made in our presence, we declare that M. Berna undoubtedly deceived himself when, on February 12 of this year, he wrote to the Royal Academy of Medicine that he could boast of affording us the personal experience of which we were in need (these are his words); when he offered to show to your delegates *con-*

clusive facts; when he affirmed that these facts were of a nature to throw light upon physiology and upon therapeutics. You have now been acquainted with these facts; you agree with us that they are by no means conclusive as to the doctrine of animal magnetism, and that they have nothing in common either with physiology or with therapeutics.

"We do not attempt to decide whether the more numerous and varied facts supplied by other magnetizers would lead to a different conclusion, but it is certain that if other magnetizers exist, they do not openly appear, and they have not ventured to challenge the sanction or reprobation of the Academy.

"(Signed)　M. M. ROUX (President), BOUILLAUD, H. CLOQUET, ÉMERY,• PELLETIER, CAVENTOU, CORNAT, OUDET, DUBOIS (Reporter).

" Paris, July 17, 1837."

When this report, taking such a decided part against animal magnetism, was read, Husson felt himself to be directly attacked, and replied. The Academy, however, accepted the conclusions of the report by an immense majority. In our opinion this report did not prove much, since general· conclusions could not be drawn from the negative experiments performed on only two somnambulists.

In order to settle the question of animal magnetism, the younger Burdin, a member of the Academy, proposed to award from his private fortune a prize of 3,000 francs to any person who could read a given writing without the aid of his eyes, and in the dark. The Academy accepted the proposal. In this way the field of ex-

periment was restricted, and it seemed that by limiting
the point at issue, it was rendered more decisive. This
was a defiance hurled by the Academy at the magnetizers, and at the first glance it might appear that
Burdin went straight to the heart of the question. He,
speaking for the Academy, seemed to say, "If there is
a single somnambulist capable of reading without using
his eyes, we will admit the existence of animal magnetism, and go into the question. If no somnambulist can
stand the test, animal magnetism has no existence." But
as Richer has observed, the dilemma is false. Somnambulists might easily be admitted to be incapable of
reading without using their eyes, and yet be genuine
somnambulists. In fact, the Academy demanded that a
miracle should be wrought before they would believe in
animal magnetism.

At this time Pigeaire, a Montpellier doctor, had a
daughter, ten or eleven years of age, who, in a state of
somnambulism, did many wonderful things, and especially
could read writing when her eyes were covered by a
bandage of black silk. This was attested by Lordat, the
Professor of Physiology at Montpellier. Pigeaire brought
his daughter to Paris, in hopes of gaining the Burdin
prize. He began with giving private *séances*, which
were completely successful; and, indeed, the private
séance generally succeeds. A very favourable report,
signed by Bousquet, Orfila, Ribes, Réveillé-Parise, etc.,
is still extant. But the scene changed when it was
necessary to appear before the commission nominated
by the Academy. The commissioners suspected that the
bandage used by Pigeaire did not serve as a complete
obstacle to the normal vision. In fact, there is nothing

apparently so simple, and in reality so difficult, as to find a bandage which is absolutely opaque; any one may see perfectly through an extremely minute hole, such as may, for instance, be perforated in a card, and especially if there are more holes than one, placed at intervals of one or two millimetres from each other. If our readers wish for further information on this interesting question, we must refer them to Déchambre's article on Mesmerism (*Dictionnaire encyclopédique des Sciences médicales*).* Déchambre took the pains to try for himself the arrangements made by magnetizers for covering the eyes of their somnambulists; and he was satisfied that none of these arrangements, although apparently very complex, would after a while prevent them from reading the writing placed under their eyes. We may add that errors become more probable from the excessive keenness of sight common in somnambulists, from the time which elapses before the reading begins, and from the contortions by which the subject tries to displace or loosen the bandage. The Academicians were, therefore, justified in rejecting the bandage used by Pigeaire. They suggested a mask or headpiece of black silk, very light ·and stretched on two iron wires, so that it might be held at the distance of six inches from the girl's face, so as not to interfere with her breathing, nor with her freedom of action. Pigeaire, on his side, objected to this, and they were unable to come to an agreement, in spite of the concessions made by the commissioners, so that the experiments did not take place. In fact, Pigeaire's stipu-

* Gerdy's paper on the same subject may also be read with interest: *Histoire académique du magnétisme animal, par Burdin jeune et Dubois d'Amiens*, p. 605.

lations would, as it was said at the time, have degraded the experiment into a mere game of blind-man's-buff.

Pigeaire was succeeded by another magnetizer, Teste, who presented himself before the Academy: he boasted of the possession of a somnambulist who could read writing which was enclosed in a box. This experiment was easily performed, and the magnetizer and the commissioners soon agreed upon the conditions. But the failure was complete, since the subject was unable to divine a single word of the writing.

The Burdin prize was not awarded.

In conclusion, Double proposed that the Academy should henceforward refuse to pay any attention to the proposals of magnetizers, and that animal magnetism should be treated as the Academy of Sciences treats the propositions which refer to perpetual motion, or to the squaring of the circle.

Such was the result of so many efforts, of such patient research, of so many discussions and reports: an absolute and complete negation of the existence of animal magnetism.

This failure of the long labours of the Academy of Medicine was, as we have already said, primarily the fault of the magnetizers. Instead of contenting themselves with the study of the simplest and most ordinary phenomena, they were bent on establishing the existence of complex psychical phenomena, such as vision by means of the occiput, or an acquaintance with future events. The Academy was also mistaken in being seduced by them into this research into the marvellous. It may be said that at the outset of the Academic history of animal magnetism, the problem was wrongly stated. It seems to us that the Academy ought to have clearly stated a question which the magnetizers were allowed to obscure; it should have been seen that among the phenomena proclaimed by the magnetizers, there might be some which were connected with known physical laws, and which might become the object of serious and fruitful study.

At any rate, the Academy ought not to have accepted Double's trenchant proposition, declaring that the question as to animal magnetism was definitively closed, as if no new facts might subsequently arise to compel the Academy to reverse its summary judgment. These new facts consist, as we are aware, in *hypnotism*, formerly regarded as an illusion, and now accepted as a truth of which no one can doubt the reality.

In fact, the history of animal magnetism is of all histories the most instructive and philosophic: we must be indeed incorrigible if it does not disgust us with *à priori* negations.

It was a matter of course that after the Academy had pronounced its sentence, somnambulists continued to see through opaque bodies, to predict future events, and to prescribe remedies, just as if the Academy had not spoken at all. Du Potet, the celebrated inventor of the magic mirror, was at this period the chief representative of magnetic science. This famous mirror, which had the effect of throwing people into convulsions, was made as follows:—The performer of the experiment described a circle on the parquet with a piece of charcoal, taking care to blacken the whole circle, and he then withdrew to a distance. The subject approached the magic circle, regarded it at first with confidence, raised his head to look at the assembly, and again looked down to his feet. "Then," says Du Potet, " the first effect might be observed. The subject drooped his head still lower with an unquiet movement of his whole person, and he revolved round the circle without losing sight of it for an instant; again he stooped lower, drew himself up, retreated a few paces, advanced anew, frowned, became gloomy, and breathed hard. The most singular and curious spectacle followed; the subject undoubtedly beheld images reflected in the mirror; his agitation and extraordinary gestures, his sobs and tears, his anger, despair, and fury—everything, in short, revealed the trouble and emotion of his mind. It was no dream nor nightmare; the apparitions were actually present. A series of events was unrolled before him, represented by signs and figures which he could understand and gloat over, sometimes joyful, sometimes gloomy, just as these representations of the future passed before his eyes. Very soon he was overcome by delirium, he wished to seize the image, and darted a ferocious glance towards it; he finally started forward to trample on the charcoal circle, the dust from it arose, and the operator approached to put an end to a drama so full of emotion and of terror."

Du Potet, a sincere enthusiast, incapable of any scientific research, explained the effects of his mirror by the intervention of magic. Gigot-Suard subsequently performed similar experiments on hypnotized subjects. This was at the time when table-turning, spirit-rapping, Home's apparitions, and other eccentricities of spiritualism were carried on. Lacordaire, in a sermon preached at Nôtre Dame in 1846, gave his adhesion to magnetism, which he regarded as the last flash of the old power, destined to confound human reason, and abase it before God; it was a phenomenon of the prophetic order.* He went on to say, "Thrown into an artificial sleep, man can see through opaque bodies, he indicates healing remedies, and appears to know things of which he was previously ignorant." Other members of the clergy went further, and practised magnetism with the avowed object of obtaining revelations from on high. The Court of Rome intervened on several occasions, and in 1856 an encyclical letter from the Holy Roman Inquisition was sent to all bishops to oppose the abuses of magnetism. The following is a translation of the Latin text:—†

"July 30, 1856

"At the general assembly of the Holy Roman Inquisition, held at the convent of Santa Maria Minerva, the cardinals and inquisitors-general against heresy throughout the Christian world, after a careful examination of all which has been reported to them by trustworthy men, touching the practice of magnetism, have resolved to address the present encyclical letter to all bishops, in order that its abuses may be repressed.

"For it is clearly established that a new species of superstition has arisen respecting magnetic phenomena, with which many persons are now concerned, not with the legitimate object of throwing light on the physical sciences, but in order to mislead men, under the belief that things hidden, remote, or still in the future may be brought to light through magnetism, and especially by the intervention of certain women who are completely under the magnetizer's control.

* *Œuvres de Lacordaire*, vol. iii. p. 246. Paris, 1861.
† Quoted by Mabru, *Les Magnétiseurs*. Paris, 1858.

"The Holy See, when consulted in special cases, has repeatedly replied by condemning as unlawful all experiments made to obtain a result which is foreign to the natural order and rules of morality, and which does not make use of lawful means. It was in such cases that it was decided, on the 21st of April, 1841, *that magnetism as set forth in this petition is not permitted.* So likewise the holy congregation thought fit to forbid the use of certain books which systematically diffuse error on this subject. But since, exclusive of special cases, it became necessary to pronounce on the practice of magnetism in general, the following rule was established on July 18, 1847:—'For the avoiding of error, of all sorcery, and of all invocation of evil spirits, whether implicit or explicit, the use of magnetism—that is, the simple act of employing physical means, not otherwise prohibited—is not morally unlawful, so long as it is for no illicit or evil object. With respect to the application of purely physical principles and means to things or results which are in reality supernatural, so as to give them a physical explanation, this is an illusion, and an heretical practice worthy of condemnation.'

"Although this decree sufficiently explains what is lawful or unlawful in the use or abuse of magnetism, human perversity is such that men who have devoted themselves to the discovery of whatever ministers to curiosity, greatly to the detriment of the salvation of souls, and even to that of civil society, boast that they have found the means of predicting and divining. Hence it follows that weak-minded women, thrown by gestures which are not always modest into a state of somnambulism, and of what is called *clairvoyance,* profess to see those things which are invisible, and claim with rash audacity the power of speaking on religious matters, of calling up the spirits of the dead, of receiving answers to their inquiries, and of discovering what is unknown or remote. They practise other superstitions of like nature, in order by this gift of divination to procure considerable gains for themselves and their masters. Whatever be the arts or illusions employed in these acts, since physical means are used to obtain unnatural results, the imposture is worthy of condemnation, since it is heretical and a scandal against the purity of morals. In order, therefore, effectually to repress so great an evil, which is most fatal to religion and to civil society, the pastoral care, vigilance, and zeal of all the bishops cannot be too earnestly invoked. Aided by divine grace, the ordinary of each diocese must do all in his power, both by the admonitions of paternal love, by severe reproaches, and, finally, by legal means, using these according to his judgment before the Lord, and taking account of the circumstances of place, of time, and persons;—he must do his utmost to avert the abuses of magnetism, and to bring it to an end, so that the Lord's flock may be preserved from the attacks of the enemy, that the faith may be maintained in its integrity, and that the faithful committed to their care may be saved from the corruption of morals.

"Given at Rome, at the Chancery of the Sacred Office of the Vatican.

"V. CARD. MACCHI.

"August 4, 1856."

It will be seen from this document that the Court of Rome appealed to a singular motive in their condemnation of magnetism. "With respect to the application of purely physical principles and means to things or results which are in reality supernatural, so as to give them a physical explanation, this is an illusion, and an heretical practice worthy of condemnation." The encyclical letter goes on to define this idea, and speaks of "weak-minded women . . . who profess to see those things which are invisible, and claim with rash audacity *the power of speaking on religious matters*, of calling up the spirits of the dead, of receiving answers to their inquiries, and of discovering what is unknown or remote." It would be impossible to declare more plainly that the Holy See proposes to maintain a monopoly of the supernatural.

Condemned by the Court of Rome, as it had been condemned by the Academy of Medicine, animal magnetism did not perish, but took refuge in the popular imagination. To this day we possess clairvoyant, and even excessively clairvoyant somnambulists who find the trade profitable. They are to be found in the drawing-rooms of private houses, as well as at public fairs. It is certain that animal magnetism will not perish, since it is one of the thousand forms assumed by that belief in the marvellous which is eternal.

As we here conclude the history of the wonders of animal magnetism, which must give place to the positive facts of hypnotism, we ought to say that it would be an error to suppose that all the phenomena of this species of legend are absolutely false. There are degrees in the marvellous. The transmission of thought, or mental suggestion, which constitutes the first stage in this domain, has been recently the subject of an article by

Ch. Richet, in the *Révue Philosophique* of December, 1884. He has attempted to show " the influence exerted in a definite direction by the thought of one individual on another in his vicinity, without any external phenomenon, appreciable by the senses." Although these phenomena are not logically connected with hypnotism, since they could be produced in Richet's friends when they were in normal health, awake, and in no sense hypnotized, yet it is true that public opinion has confounded together, under the name of animal magnetism, the nervous disturbance termed hypnotism, somnambulism, etc., and the phenomena which appear to be supernatural, such as the communication of thought, vision through an opaque body, prevision of the future, etc. For this reason we propose to say a few words on mental suggestion.

The facts in question are not absolutely new. Richet observes that we may perhaps trace the first accounts of mental suggestion to the well-known case of possession at Loudun. According to the story, Gaston d'Orléans found the Ursuline nuns agitated by frightful demoniac attacks, and he declared that they obeyed orders transmitted mentally. This was regarded as one of the chief signs of demoniac possession. De Puységur also mentions facts of mental suggestion. In the course of this century, many magnetizers have asserted that they could transmit their thoughts to somnambulist subjects; but they have been unable to prove this faculty to the satisfaction of learned bodies, which throws some doubt on their sincerity, or at any rate leads to the supposition that they unconsciously placed themselves *en rapport* with the subject by some external sign. It is now known that the slightest contact suffices to establish a communication between the one who divines and the one who suggests. Cumberland's recent experiments must not be forgotten, of which the wonderful results were shown to be explicable by very simple causes. Cumberland held the hand of an individual who had hidden, or who was thinking of, some particular object, and, with his eyes bound, went directly towards the object in question. Richet has ascertained that when the experiment succeeds, the subject, who is generally *impressionable*, unwittingly and involuntarily makes slight movements with his hand. This involuntary action betrays his thought, and puts the seeker on the right track in a way which no one who has not tried the experiment for himself would suspect.* Gley has thrown further light on Cumberland's method by his tracing of the muscular movements which explain the so-called thought-reading. The tracings clearly show that throughout the experiment there occurs in the subject's hand fibrillary contractions, slight movements of pressure, and in some cases a traction movement of the hand and whole arm. These movements increase in intensity when the object is approached, and when it is reached they suddenly cease.† Positive results were obtained from sixteen out of twenty-five persons.

We now come to Richet's experiments, and to the three orders of proof by which he sought to demonstrate mental suggestion.

1. In naming at a venture a card taken from a pack

---

* Ch. Richet, *A propos de la suggestion mentale* (*Société de Biologie,* May, 1884.)

† Gley, *Sur les mouvements musculaires inconscients en rapport avec les images* (*Société de Biologie,* July, 1884).

of playing-cards, or a picture from picture-cards, the repetition of the experiment for a given number of times will show an average more or less in agreement with the calculus of probabilities. For instance, in a hand containing six cards, the probability of guessing aright is one-sixth; that is, one time in six.

This is not the case when the card taken at random has been seen by another person; the average, varying with the sensitiveness of the subject, is then somewhat higher than that which would have been afforded by the calculus of probabilities. In 218 experiments, it would be 67 instead of 42.

2. With the aid of a rod which reveals the unconscious action of the diviner's muscles, the average is still higher than that indicated by the calculus of probabilities. The probable number in 98 experiments would be 18; the actual number was 44.

3. If the subject be placed in what are call *spiritist* conditions, which only serve to reveal the slight, unconscious movements of a sensitive person, the average obtained is very much higher than that of the calculus of probabilities.

The author considers that these latter experiments prove more than all the others. Three persons are seated at a table, engaged in conversation; the middle one, termed the medium, unconsciously moves the table, and this movement, by means of a simple arrangement, causes an electric bell to ring. Two other persons are seated at a second table, placed behind the former one, and concealed from the other three persons: one silently runs through the alphabet with a pencil; the other notes on which letter the pencil rests when the bell

4

rings.   Finally, there is a sixth person in the room, who has thought of a given word.   On consulting the letters dictated by the table, it will be seen that there is a singular correspondence between these letters and the word thought of by the sixth person, who is neither seated at the spiritist table nor before the alphabet. We give instances—

<table>
<tr><td>Words thought of.</td><td>Words dictated by table.</td></tr>
<tr><td>1. Jean Racine.</td><td>1. Igard.</td></tr>
<tr><td>2. Legros.</td><td>2. Neghn.</td></tr>
<tr><td>3. Esther.</td><td>3. Foqdem.</td></tr>
<tr><td>4. Henrietta.</td><td>4. Higiegmsd.</td></tr>
<tr><td>5. Cheuvreux.</td><td>5. Dievoreq.</td></tr>
<tr><td>6. Doremond.</td><td>6. Epjerod.</td></tr>
<tr><td>7. Chevalon.</td><td>7. Cheval.</td></tr>
<tr><td>8. Allouand.</td><td>8. Iko.</td></tr>
</table>

On a first inspection our readers will doubtless find these results very unsatisfactory.   Richet has, however, deduced some curious results from them, after submitting them to mathematical analysis.   Thus, in experiment three, where the word *Esther* was thought of, and the medium replied through the table *Foqdem*, the exactly corresponding number of letters counts as $\frac{6}{24}$ in the calculation of chances, since the alphabet consists of twenty-four letters, and the word of six letters, so that it represents six attempts to guess right.   The actual number is, however, much higher than the probable number; it is one out of six, namely—the letter *e*, which is in its right place.   On applying this analysis to all the other cases cited, Richet finds that the total probable number is equal to $\frac{57}{24} = 2$, a calculation our readers may make for themselves.   The actual number obtained was fourteen, which is very high.

Richet comes to the definitive conclusion that the probability in favour of mental suggestion may be estimated at two-thirds.   He, therefore, admits it to be probable that intellectual force is projected from the brain and echoed in the thought of another individual.   He likewise admits that this re-echo acts chiefly on the unconscious intelligence of the individual who perceives and of the individual who transmits.   This accounts for the success obtained with the spiritist table.   Under these conditions the thought of the transmitting individual acts on the unconscious thought of the medium: the latter is endowed with a faculty of semi-somnambulism, in which one portion of the brain effects certain operations without giving notice to the *ego*.   Finally, it should be said that this transmission of thought occurs in a degree which varies with the individual, since some are much more sensitive than others.

While we heartily applaud the step taken by Richet, who has had the courage to declare at his own risk what he believes to be the truth, we cannot accept his theory without reserve.   It will generally be found that the facts prove less than he asserts, and that his interpretation of them is too favourable.   One main objection consists in the fact that the calculation of chances is not adapted to decide questions of this nature: the mental transmission of thought is one of the phenomena which can only be accepted when demonstrated by proofs which should be strong in proportion as they are remote from established knowledge.   The calculation of chances is, however, for the most part incapable of affording a peremptory proof; it produces uncertainty, disquietude, and doubt.

Yet something is gained by substituting doubt for systematic denial. Richet has obtained this important result, that henceforth the possibility of mental suggestion cannot be met with contemptuous rejection.

While Richet, followed by Pierre Janet and others, has been trying experiments in France, a Society has been formed in England, called the Society for Psychical Research, which likewise makes the transmission of thought the object of study. This coincidence shows that the question is "in the air." The results obtained in England are surprising, and much higher than those of Richet. The least we can infer from them is that research should be continued in this direction, and that we should not be justified in an *à priori* denial of the possibility of these phenomena because they appear to be improbable or supernatural.

Moreover, if we consider the question of mental suggestion in its simplest aspect, if we study thought-reading in the absence of any deliberately expressed movement, we shall soon see that we touch upon phenomena which physiologists do not disdain to consider.*

Of late years Stricker has strongly insisted on the fact that a mental representation of a word or letter cannot occur without a corresponding movement in the muscles which serve for the articulation of this word or letter. This movement, constituting external speech, is

* Ch. Féré, *La question de la suggestion mentale est une question de physiologie. (Bull. Soc. Biologie,* 1886, p. 429; *Revue Philosophique,* March, 1886, p. 261.)

not generally considered as such, since it may remain unperceived by the individual in whom it occurs. Yet such a movement is visible enough to be rapidly understood by certain subjects, as we have observed for ourselves; nor will the fact appear surprising to those who understand the process by which the deaf are able to understand what is spoken. This can only be regarded as mental suggestion, since it is the reading of unexpressed ideas.

But it is not only the muscles concerned in articulation which undergo modifications of tension under the influence of external excitement, or of mental representations: all the muscles of the organism take part in this modification.* There is no paradox in the statement that certain subjects are endowed with a peculiar sensitiveness which enables them to seize these changes of form. The experiments in graphology undertaken by Richet, Ferrari, and Héricourt constitute another and no less interesting process, which shows that each psychical state corresponds to a dynamic state, characterized by objective phenomena which come within the department of physiology.

If it is true that every psychical phenomenon is accompanied by vascular modifications,† and consequently by modifications of colour, of temperature, of secretion, etc., we shall not push the hypothesis too far if we admit that excessively sensitive subjects are capable of feeling these thermic or secretory modifications.

Nothing occurs in the mind without a modification

---

* Ch. Féré, *Sensation et mouvement* (*Revue Philosophique,* October, 1885; March, July, 1886).

† Ch. Féré, *Changements de volume des membres sous l'influence des excitations périphériques et des representations mentales* (*Bull. Soc. Biol.,* 1886, p. 399).

of matter, and it is impossible to say at what point these modifications of matter may become perceptible. The study of mental suggestion is thus reduced to the reading of involuntary signs, and includes research into our most subtle reactions, and the measurement of the differential sensitiveness of various subjects, and especially of those who in their several states are hyper-excitable. This study should not be relegated to the occult sciences, to the unknowable; it is a most interesting physiological question.

CHAPTER III.

AT the time when the Paris Academy of Medicine was
condemning animal magnetism, Dr. James Braid, a
Manchester surgeon, directed the question into its proper
field—that of observation and experiment. Braid must be
regarded as the initiator of the scientific study of animal
magnetism. For this reason, since it expresses the
change of method which he effected, it is usual to substitute for that of animal magnetism the word *hypnotism*,
by which he designated the artificial nervous sleep.
Magnetism and hypnotism are fundamentally synony-
mous terms, but the first connotes a certain number
of complex and extraordinary phenomena, which have
always compromised the cause of these fruitful studies.
The term hypnotism is exclusively applied to a definite
nervous state, observable under certain conditions,
subject to general rules, produced by known and in
no sense mysterious processes, and based on modifications of the functions of the patient's nervous system.
Thus it appears that hypnotism has arisen from animal

magnetism, just as the physico-medical sciences arose from the occult sciences of the Middle Ages.

Braid began to observe the results of magnetism merely as an inquirer, and even as a sceptic. In November, 1841, he was present for the first time at some public experiments performed by Lafontaine, a Swiss magnetizer. Convinced that the phenomena which he saw were only due to an adroit imposture, he was anxious to discover by what means the operator was able to dupe his audience. He was soon satisfied that these phenomena, however strange, were quite genuine. But he saw no reason for admitting with Lafontaine that they were the consequence of the operator's personal action on his subject, by means of a magnetic fluid; he rather considered them to be due to a subjective state, independent of all external influence. This was the first result of Braid's researches; he showed that the theoretic fluid was not required to explain hypnotic phenomena.

Braid gives the following account of the way in which he arrived at this discovery. All which he saw at the first magnetic *séance* left him incredulous. At a second *séance*, six days later, his attention was struck by the fact that it was not possible for the patient to open his eyes. He regarded this incapacity as a real phenomenon, for which he sought the physical cause; it occurred to him that this cause might be found in the fixed gaze, which has the effect of exhausting and paralyzing the nervous centres of the eyes and their appendages. It signifies little whether this explanation is true or false—it is only a matter of detail; but it is important that Braid should have regarded this first symptom of hypnotism, the spasm of the *orbicularis palpebrarum*, as due to a modification of the state of the nervous system. Two days later he began, in the presence of his family and friends, a series of experiments, intended to justify his theory. He tells us that he requested his friend Walker to sit down and look fixedly at the neck of a wine-bottle, which was placed at such a height as to cause considerable fatigue to his eyes and eyelids when he looked at it attentively. In three minutes the eyelids closed, tears flowed down his cheeks, his head drooped, his countenance was slightly contracted, a sigh escaped from him, and at the same moment he fell into a deep sleep. Mrs. Braid was much astonished by the patient's fear and agitation when he awoke, for which she could see no cause, since she had not ceased to watch her husband, and she had seen that he did not approach Walker, nor touch him in any way. Braid proposed that she should herself submit to the operation, to which she readily assented, assuring those present that she should be less easily frightened than the first subject. Braid made his wife sit down and fix her eyes on the ornaments of a porcelain sugar-basin, which was placed at about the same angle with the eyes as that formed by the bottle in the previous experiment. In two minutes the expression of her features was changed; in two and a half minutes the eyelids closed with a convulsive movement, the mouth was distorted, the patient sighed deeply, the chest heaved, she fell back. It was evident that she had passed through a paroxysm of hysteria, and Braid then awoke her.

This account shows that there was nothing complex nor mysterious in the process which caused sleep; it was only necessary for the subject to concentrate his attention and his gaze for a few minutes on a given object. A *brilliant* object was sometimes employed, but this was not an indispensable condition.

From this time the reality of somnambulism was established; it became a state subject to observation, which any one could produce at pleasure. Numerous observers since Braid have repeated the experiment of the fixity of gaze, and have reproduced precisely the same phenomena. The simultaneous fixing of the attention appears to be necessary as a rule, and Braid considers that this explains why idiots cannot be hypnotized.

This important discovery throws a vivid light on religious practices which up to that time had been inexplicable. We know that Indian devotees are thrown into an ecstasy of union with God, by contemplating for hours an imaginary point in space. The monks of Mount Athos were addicted to the same practice, fixing their gaze on their navels. These are evidently hypnotic states,.produced by the fixity of gaze.

Since he showed that hypnotism could be produced by fixing the eyes on an inanimate object, such as the stopper of a bottle or the blade of a lancet, Braid proved that this nervous state did not necessarily result from the transmission of a fluid by the operator. He had therefore simplified the study of hypnotism by getting rid of all the marvellous phenomena which had discredited it for such a length of time. But Braid's conclusions were too absolute. The first conceptions of things are always simpler than the reality. It would be a mistake to suppose that the personality of the operator never has anything to do with the phenomena displayed before him. Broca's assertion must not be taken literally, "The subject is not *put* to sleep; he *goes* to sleep." The sleep produced by fixing the eyes on a brilliant object sometimes differs in certain points from the sleep produced by personal intervention. We shall soon have occasion to show that in some cases the patient displays a sort of affinity for the person who puts him to sleep, and who touches his bare hands.

Braid pursued his investigations further. His most important discovery relates to the effect produced by a given attitude on the subject's sentiments. When placed in the attitude of anger, with clenched fists, his countenance assumes a·menacing expression, and he begins to box; if he is made to imitate the action of sending a kiss, his mouth smiles. So, again, the action of climbing or swimming is produced when the body is placed in the position required for executing the several acts.

These were Braid's two chief discoveries; he also made several observations of which the justice has now been admitted. He ascertained that the character of the sleep was not always the same, but that it consisted of a series of states, varying from a light slumber up to the most profound sleep. He observed that breathing on the face had the singular effect of changing the hypnotic state, and breathing on it for the second time caused the subject to awake. He also observed that the senses, especially those of touch, smell, and hearing, might suddenly become excessively acute in hypnotized subjects, and it appeared to him that this sensorial modification might afford a rational explanation of some of the marvellous effects obtained by professional magnetizers. Finally, he observed that verbal suggestion might produce hallucinations, emotions, paralysis, etc. Suggestion during the waking state, which has latterly been asserted by some writers to be possible, did not escape his notice.

Although so many of his observations were just, Braid's descriptions of hypnotism are not definite; they contain an indiscriminate account of all the symptoms of hypnotism, anæsthesia, hyperæsthesia, hallucinations, paralysis, suggestions of theft and other criminal acts, unilateral hypnosis, duplication of the consciousness, etc., as if all these phenomena had not their peculiar conditions, and did not belong to distinct states. Braid's imperfect work has been completed by the Salpêtrière school, which shows that hypnotism is a nervous condition, presenting characteristics which vary in intensity, if not in their nature, so that it is possible to distinguish the several phases or states in which the action of the subject varies.

In addition to the want of classification betrayed by this disorderly exposition of facts, Braid has erred in putting in one category the unproved and the uncertain, the uncertain and the purely imaginative. A few pages of his book suffice to show that we have to do with a believer rather than with an observer.

Braid has also been blamed for his unsatisfactory experiments in *phreno-hypnotism*, intended to prove the possibility of exciting special sentiments, ideas, and acts, by pressing on the *bumps* of the skull of a hypnotized subject. The account of these experiments occupies an important place in his *Neurypnology*. Braid, after taking care to inform us that, while making use of phrenology, he is no materialist, confidently asserts that he could inspire the idea of theft by pressing on the

organ of acquisitiveness; of fighting, by pressing on that
of combativeness; of prayer, by pressing on the organ of
veneration, etc.

The following experiment was the most curious of
the series, and will give an idea of the others. Acquisitiveness was excited, and the subject stole a silver snuff-
box from one of the spectators; the pressure was then
transferred to the organ of conscientiousness, and the
patient surrendered the object with a striking air of
contrition. Braid seems to have foreseen the charge of
simulation, and he takes care to affirm that several of
his phrenological experiments were performed on persons
who knew nothing about phrenology, and whose honour
was unimpeachable. It is easy, up to a certain point, to
understand the strange illusion of which Braid was the
dupe. He had not observed the importance of that
frequent source of error called *unconscious suggestion*.
It is now known that an indiscreet word uttered before
subjects very sensitive to suggestion is enough to show
what is expected of them, and to make them act in the
sense intended by the operator. A gesture may sometimes produce the same effect, and this explains how, in
some public exhibitions, the magnetizer, having agreed
with his subject to deceive the spectators, is able to
make him obey mental orders without expressing them
verbally. There is in reality no communication by
thought, but by signs which are comprehended by the
subject with extraordinary quickness of perception. In
Braid's experiments it is probable that something
analogous occurred, although there was no imposture.
Braid was doubtless as honest as his subjects, but the
latter unconsciously obeyed a gesture or word, or were

unconsciously influenced by the recollection of a previous *séance.* This seems the more probable assumption, since Braid's subjects were often people in good society, assembled to take part in a *séance* of phreno-hypnotism, and who, after seeing what Braid effected on others, voluntarily submitted to be the subjects of experiment.

Braid's errors are not, however, wholly devoid of truth. Numerous observers have declared that pressure on the heads of hypnotic subjects produces a surprising variety of sensory and motor effects.

As a physician, Braid was much occupied in applying hypnotism to therapeutics. His observations refer to diseases of the eye, to tic-douloureux, nervous headaches, spinal irritation, neuralgia of the heart, palpitations and irregular action of the heart, epilepsy, paralysis, convulsions, tonic spasms, affections of the skin, rheumatism, etc. We cannot refrain from the belief that here again Braid was deceived in more than one instance, but he must be credited with having made a fairly methodical study of hypnotic therapeutics.

The results of Braid's labours have in our day been considerable. He has the merit of having proved that animal magnetism is a natural phenomenon, a definite nervous condition, produced by means of known processes. Lasègue regards him as an indifferent physiologist. But this matters little, since many more intelligent and liberal minds have not the merit of having discovered a single new fact. Indeed, it appears that a certain narrowness of mind, allied with an obstinate will, is to some extent characteristic of the innovator.*

* We subjoin a list of Braid's works : *Neurypnology; or, The Rationale of Nervous Sleep, considered in relation with Animal Magnetism,* by James

Braid's discovery had little success in his own country, although it obtained the support of the physiologist Carpenter. In 1842 he submitted his researches to the medical section of the British Association, and offered to repeat his experiments before a special commission. The offer was formally rejected, and the section proceeded to other matters. It was said that this subject, like so many others, must make its way independently of learned bodies. Braid was not · discouraged, and became the propagator of hypnotism with the indefatigable ardour which is characteristic of innovators, and which we have lately observed in Burq, the inventor of *metallo-therapia.* He held many experimental *séances* in London, Liverpool, and Manchester, without obtaining the justice due to him.

Braid's theory had more success in America, but not under his own name. In 1848 an American named Grimes, who does not appear to have been acquainted with Braid's discovery, showed that most of the hypnotic phenomena could in certain subjects be produced in the waking state by means of verbal suggestion. This theory, which passed in the United States under the somewhat absurd name of *electro-biology,* reached England in 1850, and produced a new movement in favour of hypnotism.

Although extracts from Braid's works were published by Littré and Robin, by Robin and Béraud, etc., and

Braid (London : John Churchill, 1843); *The Power of the Mind over the Body* (1846); *Observations on Trance, or Human Hybernation* (1850); *Magic, Witchcraft, Animal Magnetism, Hypnotism, and Electro-biology* (1852); *The Physiology of Fascination* (1855); *Observations on the Nature and Treatment of Certain Forms of Paralysis* (1855).

there was an article on the subject in the *Presse* by Meunier, his theories were little known in France.

In 1850 the question was, however, again brought before the French public by Azam, a Bordeaux surgeon. Azam had been called in to see a poor girl who was said to be insane, and who presented the singular phenomena of spontaneous catalepsy, of anæsthesia, and of hyperæsthesia. Azam was acquainted with the magnetic phenomena of artificial somnambulism, and was struck by the correspondence between these and those which occurred spontaneously in his patient. One of his colleagues mentioned to him Braid's experiments, which were reported in Todd's *Encyclopedia*, and he tried to repeat these experiments on his patient, not without misgivings. He tells us that "at the first attempt, after being subjected for one or two minutes to the usual process, the patient fell asleep: the anæsthesia was complete, and it was evident that she was in a state of catalepsy. Hyperæsthesia afterwards supervened, accompanied by the power of answering questions, and other symptoms indicative of the exercise of the intelligence." * Similar experiments were successfully performed by Azam on another girl living in the same house, for the most part such experiments as had been described by Braid. We quote an instance of suggestions by means of the muscular sense: " If, during the period of catalepsy, I place Mlle. X——'s arms in the position of prayer, and leave them thus for a certain time, she states that her thoughts are fixed on prayer, and that she supposes herself to be present at a religious rite. When placed with folded arms and drooping head, she

* *Archives de Médecine*, 1860, p. 8.

feels her mind possessed by a series of ideas of humility and contrition. When her head is raised her ideas become haughty." The hyperæsthesia of the senses is no less decided. Azam asserts that the hearing becomes so acute as to distinguish the ticking of a watch at the distance of nine or ten yards: this sensitiveness to noise fatigues the subjects, and an expression of pain passes over the face at the rolling of carriages, the human voice, etc. When a bare hand is placed behind her back, at a distance of forty centimetres, Mlle. X—— stoops forward and complains of feeling the heat.

Azam was, however, chiefly struck by the general anæsthesia which frequently accompanied the hypnotic sleep. In concert with Broca, he sought in hypnotism a fresh mode of producing anæsthesia during surgical operations. This idea gained ground. Broca remarks that a method which introduces no foreign substance into the system appears to him to be absolutely inoffensive. This, however, is erroneous, since death may be produced by a suggestion. In 1851 Broca and Follin put a woman under hypnotism before making an incision in an abscess in the anus. This fact was communicated to the Academy of Sciences by Velpeau, who, in announcing with satisfaction this "new discovery," appeared to have no doubt that animal magnetism, which had been condemned by the Academy, had reappeared under a new name. A few days later Guérineau, of Poitiers, employed the same hypnotic anæsthesia during the amputation of a thigh. The interest in hypnotism became general, and it was remembered that as early as 1829 Cloquet had amputated the breast of a magnetized woman, and that Loysel had performed very serious operations under like conditions. It was, however, a transient interest, since the surgeons perceived that the hypnotic sleep could not be produced in all subjects ; that even in those most susceptible to it, a series of daily hypnotizations must precede the operation, and that sometimes, instead of producing anæsthesia, the converse effect of hyperæsthesia was produced. These failures were partly due to the fact that it was not then known that suggestion might be used to produce insensibility. Chloroform was, therefore, soon preferred to hypnotism as the safer and more convenient means. The year 1860 witnessed the dawn and decline of the prevailing fashion of employing hypnotism to produce surgical anæsthesia.

The question of animal magnetism, which had been proscribed twenty years before by the Academy of Medicine, was, however, reopened. The reality of the nervous sleep was no longer disputed; the mode of producing it was known, as well as its main symptoms. Distinguished physicians were now anxious to study these phenomena, without fear of compromising themselves. It was at this time that the works of Demarquay, and Giraud-Teulon, Gigot-Suard, Liébault, and Philips (Durand de Gros) appeared.

The chief result of these researches was to confirm Braid's work in essential particulars. It was again proved that the personality of the hypnotizer is not a necessary element in producing the subject's sleep. Demarquay and Giraud-Teulon, in order to ward off the influence of the experimenter's gaze, made use of a polished steel ball, which was mounted on a stalk and fastened to a diadem ; this diadem was placed on the subject's head, and his eyes were consequently drawn into the indicated convergence without the intervention of the experimenter.* It is needless to add that this method produced sleep in the subject, just as other methods did. Gigot-Suard even ascertained that a brilliant object need not be presented to the eyes, and that the fixity of gaze would suffice. It was enough to order the subject to look at his nose, and then immediately to bandage his eyes. This also produced hypnosis.

Demarquay and Giraud-Teulon agreed that a predisposition to hysteria was a general condition of hypnotic effects. In fact, results were only obtained from four persons out of eighteen, and these four were all women; the men submitted to experiment were altogether refractory. Moreover, in one of these women the attempt at hypnotization produced the first symptoms of an hysteric attack. Hence they concluded that the nervous state designated as hypnotism was not physiological, but altogether morbid. The work by Demarquay and Giraud-Teulon is brief, accurate, and full of carefully observed facts, without the mystical tendency which is found in Braid. It is perhaps the first work on hypnotism of a strictly scientific character.

Durand de Gros, better known as Dr. Philips, since he was one of the proscribed of December 2, and assumed this name in order to return to France, delivered public lectures on hypnotism in Belgium, Switzerland, France, and Algeria. In 1860 he published a *Cours théorique et pratique de Braidisme*, in which he developed his ideas on the mechanism of hypnosis. But the medical world was not much moved by his abstract conceptions

* *Recherche sur l'Hypnotisme* (*Gazette Médicale de Paris*, 1859, 1860).

of the *hypotaxic state* and of *ideoplasticism*, of which we will only say a few words. According to this author, the exercise of thought is necessary for the regular diffusion of nervous force into the sensory nerves ; the exercise of this mental activity is suspended by hypnotism, or rather is reduced to a minimum by submitting it to the exclusive excitement of a simple, homogeneous, and continuous sensation. Since the nervous force is no longer consumed by thought, it accumulates in the brain, and this sort of nervous congestion is termed the *hypotaxic state.* But, by a special impression on the sight, the hearing, or the touch, a given point of the brain may be excited, so that all the disposable nervous force may be accumulated on it. The same result may be obtained with a mental impression. as with a sensorial impression ; it awakens the activity proper to a given part of the brain, and produces the most varied effects. This is *ideoplasticism.*

Durand de Gros's theories somewhat resemble those set forth five years later by Liébault, a physician of Nancy, in a work entitled, "*Sleep, and the states analogous to it, specially considered in the action of the morale on the physique*" (Nancy, 1866). In his preface Liébault writes: "In my endeavour to study the passive modes of existence, I have first sought to demonstrate the truth that they are the effects of a mental action, and then to make my readers acquainted with their properties, from the point of view of the action of the *morale* on the *physique.*" In these words we find the germ of the idea developed by subsequent writers, who wish to prove that all the phenomena of artificial sleep,—both mental and physical phenomena, such as contractions, catalepsy, etc.. —are produced by *suggestion.* Thus, Liébault asserted that artificial as well as natural sleep was produced by an act of the intelligence, that is, by concentrating the attention on one idea, that of going to sleep. This explanation does not apply to those persons who are hypnotized against their will. Liébault goes further, and maintains that modifications of the attention, its too energetic retreat into the brain, etc., cause the difficulty of breathing, the dilatation of the pupils, the weight of the head, singing in the ears, cyanosis, and the palpitations of the heart which accompany the approach of sleep. In Liébault's opinion, attention appears to sum up the action of the mind on the physique. Concentration of the attention causes the isolation of the senses, the cessation of muscular movements, the establishment of a *rapport* between the somnambulist and the operator, catalepsy, etc. The afflux of attention to the organs of the senses increases their power of perception; its accumulation on the "*empreintes sensorielles*" quickens the memory, and so it is with the other senses. On waking from a state of profound hypnotism, there is oblivion, which is due to the fact that all the nervous force accumulated in the brain during sleep is, on awaking, again diffused throughout the organism; since the nervous force is diminished in the brain, it is impossible for the subject to recall to mind that of which he was previously aware.

Liébault's ideas were received with incredulity; his mode of practice appeared to be so singular that it was rejected by his colleagues without further examination. He lived in retirement, apart from the medical world, and entirely devoted to his convictions and to his patients, who were almost wholly of the poorer class. It is not difficult to understand the cause of his failure. His book does not contain any clear and definite account of hypnotism ; the symptoms which result from this profound modification of the system are not the object of a methodical study, and his descriptions are vague and without definite character. There is, perhaps, not a single scientific proof of hypnotism in the whole book. Yet we must give Liébault credit for having been a conscientious observer, convinced of the truth of his practice. It is said that his convictions brought him into unpleasant relations with his colleagues, and it is probable that they would never have been accepted without the labours of Charcot and his pupils, who re-established the study of hypnotism, simply by giving an accurate description of the physical characteristics of some of the nervous states designated by that name.

The theories of Braid were now again in the ascendant. Up to 1878 nothing of much novelty was contributed to them. We need only mention the works of Mesnet (1860), of Lasègue (1865), of Baillif (1868), of Pau de Saint-Martin (1869). No advance was made, but the same ground was traversed again.

There is a good account of the works of this period in an article by Duval, which appeared in 1874 in the *Dictionnaire pratique de médicine et de chirurgie*. At the same date, Dechambre declared in the *Dictionnaire encyclopédique des sciences médicales* that animal magnetism did not exist.

By degrees the question sank into silence and oblivion. The more earnest minds turned away from it, and abandoned the subject to professional magnetizers, who contrived to make money by public exhibitions of hypnotism.   From time to time a man of note attempted to shake off the general indifference, but no response was made. In 1875, Ch. Richet published in the *Journal de l'anatomie et de la physiologie* the result of some researches on hypnotism, which he had made while house-surgeon of a hospital. Although the paper was interesting and full of facts, it obtained little notice.

At about the same time the somnambulism of animals was studied in Germany. As Richet justly observes: " In order to judge of the question of simulation, nothing can be simpler than to perform experiments on beings incapable of playing a part."   But it was ascertained, on setting to work, that the symptoms of somnambulism in animals are by no means strongly marked.

As early as 1646, Father Athanasius Kircher relates, in a book entitled *Ars magna lucis et umbrae*, that if a cock, with his legs tied together, be placed before a line made upon the floor with white chalk, he becomes at the end of a few moments perfectly motionless ; if the string be untied and he is excited, he does not issue from the cataleptic state. This experiment may be of still earlier date, since it has been ascribed to Daniel Schwenter (1636). However this may be, in many countries the hypnotization of poultry became a popular amusement. In 1872, Czermak carefully repeated these experiments ; he hypnotized a cock without making use of the ligature, or of the chalk line, and kept the animal immovable. He extended the experiment to other animals, to sparrows, pigeons, rabbits, salamanders, and crabs.* Preyer,† whose treatise on the subject is the most

---

* *Comptes rendus de l'Académie de Vienne*, 1872, p. 361.
† *Die Kataplexie*, etc.   Jena, 1878.

complete which we possess, ascribed most of the phenomena observed under these conditions to fear. This author holds that strong excitement produces the cataleptic state, that is, a paralysis due to fear. For instance, if a lizard's tail, or a frog's foot, is suddenly pinched, the animal becomes petrified, sometimes for several minutes, and is incapable of moving its limbs. Gentle and protracted excitement is needed to effect the hypnosis of animals. If the nostrils of a guinea-pig are kept for some time slightly compressed with a pair of pincers, the animal becomes hypnotic, and is thrown into such a stupor that it can be placed in the most absurd positions without being awakened. , This arbitrary distinction between catalepsy and hypnotism has not been accepted. We need only note that many animals can be hypnotized, either by a brief and strong excitement of the skin, or by a repeated and fainter action of the same kind.

The experiments on the frog are interesting, and easy to reproduce. Heubel * has shown that if a lively frog is lightly held between the fingers, with the thumb on the belly, and the four fingers on the back, the animal becomes perfectly motionless at the end of two or three minutes; it may be stretched upon its back, or placed in all sorts of positions, without making any attempt at defence or escape. The same paralytic state may be produced by gently scratching the frog's back. But it must be admitted that none of these facts throw much fresh light on animal magnetism, and we do not, therefore, insist on them further.

We now come to the year 1878, and to the researches of the Salpêtrière school.

* *Archives Pflüger*, vol. xiv.

The history of animal magnetism has shown that if, up to late years, the existence of the nervous sleep, and of the various phenomena allied with it, has been doubted, it is chiefly because the experimenters wanted method, and were principally concerned with the study of complex psychical phenomena. Such phenomena often lack the material characteristics which would place them beyond dispute. Since the proofs of these remarkable manifestations were wanting, it was at once concluded that they were, at any rate, hypothetical, if not false.

The disputes and doubts might have gone on indefinitely, but for the intervention of material facts, which it was impossible to interpret in different senses. These material facts could not be at once discovered in the domain of the complex phenomena which had attracted the attention of the early experimenters; they belonged to the purely physical order of things.

We must add that these physical signs of hypnosis have not hitherto been observed in their complete development, except in subjects affected by hysteria. Hence it follows that the hypnotism which first took its place in science is that of hysterical patients, and it is still termed *profound hypnotism*, both to characterize the intensity of its symptoms, and to distinguish it from the feebler forms which had, up to that time, been exclusively studied by physicians, and which may now be grouped under the name of *slight hypnotism*.

The method which led to the revival of hypnotism may be summed up in these words: the production of material symptoms, which give to some extent an anatomical demonstration of the reality of a special state of the nervous system. This is merely an application of

Descartes's rule, that we should go from the simple to the compound. Before adopting this method, we have passed through an age of senseless errors and sterile discussions.

It is to Charcot that the honour must be assigned of having been the first to enter on this course, in which he has been followed by numerous observers. The violence with which he was attacked is a proof of the important part he took in the question. Whatever objections may be made to his description of the different states known under the name of hypnotism, it is certain that the application of the nosographic method to this study enabled Charcot to establish phenomena within the domain of science which had hitherto been regarded as beyond its range. Charcot was not only fortunate enough to establish the scientific value of hypnotism, but to obtain compensation for his earlier academic failures by his triumphant readmission into the Academy of Sciences.*

The researches of the Salpêtrière school served as the point of departure for a fresh scientific movement, which continues up to the present day.

In 1880, Heidenhain, an eminent German physiologist, resumed the study of hypnotism, prompted by some public performances at Breslau, given by a Danish magnetizer named Hansen. Heidenhain's paper † gave the signal for several other German publications, among

* J. M. Charcot, *Essai d'une distinction nosographique des divers états compris sous le nom d'Hypnotisme* (C. R., Ac. des Sciences, 1882).

† Heidenhain, *Der sogenannte thierische Magnetismus; Physiologische Beobachtungen* (Leipzig, 1880); Heidenhain und Grützner, *Halbseitige Hypnotismus, Hypnotische Aphasie, Farbenblindheit, u. Mangel des Temperaturismus bei Hypnotischen,* in *Bresl. Artzl. Zeitschr.,* ii. 4. 1880.

which we may mention that by Grützner,* by Berger,†
by Baumler, ‡ by Preyer, § by Schneider. ‖ In France
we find, among others, P. Richer, Bourneville and
Regnard, Dumontpallier and his pupils, Ladame, Bottey,
Pitres, Brémaud, Bernheim, Beaunis; in Italy, Tamburini
and Seppili, and Lombroso; in England, Hack Tuke.

* Grützner, *Ueber d. neureren Erfahrungen aus dem Gebiete des soge-
nannten thierischen Magnetismus (Cent. f. Nerv. Psch.*, 10. 1880).

† Berger, *Hypnotische Zustände, und ihre Genesung*, in *Bresl. Arztl.
Zeitschr.*, ii., 10, 11, 12. 1880; *Das Verhalten der Sinnesorgane, in hypno-
tischen Zustand*, in *Bresl. Arztl. Zeitschv.*, iii. 7. 1881; *Experimentelle
Katalepsie; Deutsch. med. Wochenschrift*, vi., 10. 1880.

‡ Baumler, *Der sogen. animalische Magnetismus, oder Hypnotismus.*
Leipzig, 1881.

§ Preyer, *Die Entdeckung des Hypnotismus.* Berlin, 1881.

‖ Schneider, *Diepsych. Ursache der Hypnot. Erschein.* Leipzig, 1880.

CHAPTER IV.

THE MODES OF PRODUCING HYPNOSIS.

As far as its mode of production is concerned, hypnotic sleep does not essentially differ from natural sleep, of which it is in fact only a modification, and all the causes which produce fatigue are capable of producing hypnosis in those who are subject to it; it is in this sense that we may say with Richer, that all means are effectual, if only they are applied to a predisposed organism.

Sensorial excitements produce hypnosis in two ways: when they are strong and abrupt, or when they are faint and continued for a prolonged period.

The former mode of excitement was studied for the first time by Charcot and his pupils, who employed, among other means, vivid impressions on the sight, such as the sudden introduction of a solar lamp into a dark room, fixing the eyes on the sun, the incandescence of a strip of magnesium, the electric light, etc. In hysterical subjects the intense excitement immediately produces catalepsy. If the patient is seated at work, is standing, or walking, she is transfixed in the attitude in which she was surprised, and fear is expressed in her countenance and in her gestures. The same effect may be produced by an intense noise, like that of a Chinese gong, by a whistle, or by the vibration of a tuning-fork. When the subject is predisposed, comparatively slight, but unexpected noises, such for instance as the crackling of a piece of paper, or the chinking of a glass, are enough to produce catalepsy.

If the excitement is moderate, rather than violent, it must be prolonged in order to cause the hypnotic sleep, which, however, it scarcely ever fails to produce. The subject is put to sleep after Braid's method, by fixing his gaze for a few moments on an object which may be slightly luminous, or altogether dark, such as a black stick, which should be held near the eyes and a little above them, so as to produce a convergent and superior strabismus. After a while the eyes become humid and brilliant, the gaze becomes fixed, the pupils are dilated. When the object is withdrawn, the subject remains in a cataleptic state; if it is not withdrawn, the subject soon falls backwards with a sigh, there is a slight frothing on the lips, and lethargy ensues. The convergence of the eyes alone will produce sleep, as for instance at night (Carpenter); some subjects fall asleep spontaneously when their eyes are fixed upon their needle-work, when they are reading, or looking in the mirror while dressing. Monotonous sounds also produce sleep. Weinhold and Heidenhain produced hypnosis by causing the subject to listen to the ticking of a watch; and a faint but continuous musical sound may produce the same effect. It is also well known that monotonous action on the hearing, a nurse's lullaby, the noise of the wind, the reciting of prayers, have a marked effect in producing natural sleep in many people.

It likewise occurred to us to produce a lethargic

sleep by fatiguing the sense of smell with a protracted odour of musk. No experiments have been made on the sense of taste. Tickling of the pharynx has succeeded with many subjects, but in this case it may be from complex reasons, since the subject nearly always fixes his eyes and keeps the thorax motionless.

Some facts appear to indicate that an excitement of the organs of the senses which does not act upon their special functions, but only mechanically, may produce like effects. Thus, when the eyeball is compressed through the closed lid, which was often done by Lasègue, hypnosis may be produced in some subjects, and a like effect may be produced by pressure on the external meatus of the ear. These modes of hypnotization belong, as we think, to the group of those which act by exhaustion of the special senses. In fact, a pressure on the eyeball, however slight, produces irritation at the base of the eye, whence there follows a sensation of light. When the external orifice of the ear is compressed, there is a pressure on the membrane of the tympanum by means of the air contained in the tube, and it may easily be shown that this causes a continuous murmur, which fatigues the sense of hearing, so that in this case also sleep results from exhaustion.

The hypnotizing processes in which a method involving contact with the skin is necessary, are, however, susceptible of more than one interpretation. We know that magnetizers formerly made use of what are termed *passes;* these passes consist in lightly touching the subject, either directly, or indirectly, through his clothes, and a prolonged repetition of these gestures produces sleep. Ch. Richet has ascertained that a gentle excitement of the skin may produce the somnambulist sleep as well as the excitement of the special senses ; yet it may be assumed that the success of the passes is greatly due to the psychical element.

We must note one interesting point in the history of hypnotizing processes, by means of irritation of the skin. On looking over the writings of magnetizers in the first half of the century, we are struck by the recurrence of certain gestures which contributed to discredit animal magnetism. It appears that the experimenter often caused his subject to sit down opposite to him, pressed his or her knees within his own, grasped the thumbs with his hands, and sometimes applied his forehead to that of the subject of experiment. These gestures, which appeared to be indecent, and unnecessary for the purpose he wished to effect, were in fact founded on accurate observations, which have since been verified.

It has been ascertained that when the scalp of hypnotized subjects is slightly irritated, the character of the sleep is changed. Thus, individuals plunged in the state designated by Charcot as lethargic or cataleptic, may be made to pass into the somnambulist state by a slight friction in the region of the scalp. Heidenhain, Grützner, and Berger, by slight and prolonged friction on one side of the heads of subjects in the waking state, have produced in them a *unilateral hypnosis*, displayed by an excessive muscular excitability. The influence of irritation localized in certain regions has recently been well described by Pitres, who has shown that in some subjects there are zones he terms hypnogenic, sometimes superficial, sometimes deeply seated, and that even a slight irritation of these zones may produce hypnosis, or occasionally cause it to cease.* Such zones may be found in all parts of the body, but most frequently in the vicinity of the joints, on the scalp, and especially on the forehead, and also at the root of the thumb. The legitimate observation of facts therefore justifies the gestures formerly in use, and we must not hastily condemn or deny that which we do not understand.

Irritation of the skin is as effectual when it is done with a feather, or some other inert body, as with the hand. We have ascertained that the sleep may be produced in several instances by placing a magnet close to an hypnogenic zone. We have also observed that the subject may put himself to sleep by pressing on such a zone. It should be added that each subject may display different hypnotic zones, not only as to their site, but as to their action; lethargy, catalepsy, and somnambulism may result in their several forms from the excitement of one or other of these zones.

Heat may produce the same effect as a mechanical excitement of the skin. Berger showed that he could produce hypnosis by holding his hot hands near the head of a person in a natural sleep ; the heat disengaged from his hands produced this effect, for when he wore woollen gloves, or covered the sleeper's head, hypnotization did not occur. Berger also obtained like effects by placing metal plates, moderately heated, near the heads of his subjects.

An excitement which is not felt may have a hypnotizing effect, since consciousness is a super-added element, which is not essential. Thus the magnet, which acts as a peripheral excitement, may hypnotize a subject

* A. Pitres, *Des zones hystérogènes et hypnogènes.* Bordeaux, 1885.

without his perceiving the action exerted by this body on his organism. The influence of the magnet on hypnosis was first pointed out by Landouzy, in 1879, and the fact was afterwards verified by Chambard, and by the present writers.

Hypnotization by sensorial excitement, or by a physiological process, may be summed up as follows :—

1. By excitement of the sense of sight : (*a*) Strong and sudden excitement, by luminous rays, by solar or electric light, or by the sudden incandescence of a magnesium wire ; (*b*) slight and prolonged excitement, by fixing the eyes on an object, brilliant or otherwise, which is placed near the eyes, and somewhat above their level.

2. By excitement of the sense of hearing : (*a*) Strong and sudden excitement, by a gong, by copper instruments, etc. ; (*b*) slight and prolonged excitement, by the ticking of a watch, the vibrations of a tuning-fork, or any other monotonous sound.

3. By excitement of the senses of taste and smell.

4. By excitement of the sense of touch : (*a*) Strong and sudden excitement, by pressure on the hypnogenic zones ; (*b*) slight and prolonged excitement, by passes, contact, action of heat or of the magnet.

These several physiological processes act very differently on different subjects. When used in combination, their effect may be greater or more rapid. Although, as Braid has shown, the operator's personality has not the importance which was formerly ascribed to it, yet it cannot be said to be altogether negative. It can easily be proved that some experimenters are more successful than others, at any rate with some subjects. This elective phenomenon is not unimportant, and is perhaps partly due to the specific heat, smell, etc.

If this elective affinity exists in physiological processes, it is much more manifest in those processes which are psychical. In fact, hypnosis is not produced only by sensorial and peripheral excitement; it is also effected by central excitement, that is, by acting on the imagination. It may be asserted that, whenever the subject is warned that he is about to be hypnotized, his mind contributes to the success of the operation, and the sleep is partly due to psychical action.

The Abbé Faria, who induced sleep by *intimation*, has clearly shown that hypnosis may be effected by psychical action. His process consisted in desiring the subject, in an imperious voice, to go to sleep, and sometimes, without uttering a word, a commanding gesture was enough to effect his purpose. Faria's simple process is rarely employed, and *insinuation* is often substituted for intimation. 'Sleep may, for instance, be induced by telling the subject that he is sleepy or heavy, that his eyelids are closing, that he does not hear, nor see, etc., or —as we have ourselves observed—when the experimenter himself feigns to sleep. This gentle process is perfectly successful with subjects who have previously been hypnotized in other ways, and it succeeds at once with predisposed subjects, who have been under a course of treatment, and who feel confidence in the operator, and in the result of the operation. It is, in fact, only suggestion in the waking state.

This suggestion is often veiled by manœuvres which formerly led to the belief that it was possible to magnetize from a distance. A susceptible subject could be put to sleep by making passes through a door, if only the subject was aware that a magnetizer was present with that purpose in view. This experiment, intended to show that somnambulism is produced by a fluid which escapes from the magnetizer's body, and passes through opaque bodies, simply proves that the subject's fixed idea that he is being hypnotized is enough to put him to sleep, and this is a psychical impression. In this way it can be explained how a magnetizer in Paris can hypnotize one of his subjects in the country, when the latter is aware on what day and at what hour the operation is to begin; and, again, how some subjects are hypnotized by causing them to touch objects to which magnetic virtue has been openly ascribed. This likewise explains the action of magnetized water and magnetized trees. But the most striking experiment is the suggestion of sleep after a long interval of time. The subject is assured, with the necessary firmness and authority, that after so many days, at such an hour, he will spontaneously fall asleep. On the day appointed and at the given hour the suggestion is realized; the subject is overcome by sleep in the midst of his occupations, and in whatever place he may happen to be.

Several writers, who have observed the power of suggestion as an hypnogenic agent, have regarded it as universally present. Thus Braid asserts that the imagination of the subject is an indispensable element in the success of the experiment; he declares that the most expert hypnotizer will exert himself in vain, unless the subject is aware of what is passing and surrenders himself, body and soul. In our day, some authors have maintained that the expectant attitude was the cause of all hypnotic phenomena, as well as of the phenomena of metallotherapeutics. Schneider and Berger consider that hypnosis is produced by a unilateral concentration of the attention. These assertions are too absolute. A whole series of purely physical agents exist, which prove that sleep can be induced without the aid of the subject's imagination, against his will, and without his knowledge.

We will mention, in conclusion, some of the experiments made by one of the present writers,* which confirm the idea already suggested by Braid, that hypnosis results from the exhaustion of the cerebral influx. An experimental proof can be given that all the sensorial excitements which induce hypnosis act by exhaustion, for the first effect of these excitements is an exaggeration of the motor phenomena. If the subject is made to hold a dynamograph in his right hand in such a way as to exert no pressure on it, and if he is then hypnotized, it can be ascertained that a motor discharge occurs in the interval between the excitement and the sleep. There is an intense pressure of the fingers on the dynamograph, and, indeed, the movement extends to all the muscles of the body. It is therefore probable that the hypnogenic excitement provokes an exhausting activity.

We are here met by the difficulty that the theory of exhaustion does not explain the sleep produced by suggestion. It has often been said that the psychical element in hypnosis vitiates all the attempts to give a physical explanation of this state. While admitting that the problem is difficult, we think it possible to

* Ch. Féré, *Inhibition et épuisement* (*Soc. de Biologie*, May 7, 1886); *Impuissance et pessimisme* (*Revue Philosophique*, July, 1886); *La médicine d'imagination* (*Progrès Medicale*, 1886, p. 717).

reconcile some psychical processes of hypnotization with processes due to exhaustion. All kinds of suggestion consist in making one idea predominant in the subject's mind; the suggestion of sleep is included in this category, and hypnotization is effected by the idea of sleep. Repeated experiments, which we shall afterwards mention in detail, show that every idea is an image, that every image recalls an anterior sensation. From this point of view, hypnotization by suggestion consists in hypnotization by physical excitements, not actually occurring, but remembered. In confirmation of this assertion, we give an example of an experiment communicated to us by Ballet. The suggestion was made to a subject, either in her waking state or in a previous sleep, of an electric lamp, shining from the corner of the room. The subject was awake and conversing tranquilly. When told to look in the corner where the imaginary lamp was placed, she was at once attacked by catalepsy, just as if the electric ray had shone upon her face. Hallucination, that is, the image of the luminous impression, produced the same effect as the actual impression, because it was recalled to her mind. So it seems probable that the suggestion of sleep only effects its purpose by inducing the recollection of certain impressions of fatigue which involve exhaustion in the same way as a physical excitement.

The awakening of the hypnotized subject, as well as his hypnotization, may be effected by two different processes—by a peripheral impression, or by a central and psychical impression. It is generally enough, in order to awaken the subject, to breathe lightly on his eyes or forehead. The wind from a pair of bellows may be substituted without inconvenience for breathing from the mouth, or a few drops of water may be sprinkled on the face. When these means fail, the subject's eyelids are opened, in order to breathe strongly on the corneæ. And, in the case of hysterical patients, who do not awake under this treatment, pressure is applied in the region of the ovarium. Pitres has also shown the existence of superficial zones in many hysterical subjects, which may be excited in order to awaken them. It is very probable that they might be awakened by addressing special senses, particularly those of sight and hearing. But nothing certain is known on these points. If the experimenter breathes on one half of the forehead, while sheltering the other half with a screen, only half of the body is awakened. The subject may also be awakened by a psychical impression. When the order to awake is repeated a certain number of times, the subject awakes, just as he goes to sleep when ordered to do so.

We see that there is a certain parallelism between the causes which produce hypnotism and those which remove it, and that in both cases it is done by excitement, whether of the surface of the skin, or of the special senses, or by a psychical excitement. This relation between the two processes is still more marked in some hysterical patients in whom there are found zones endowed with inverse properties, at once hypnogenic and the reverse. When the patients in question are awakened, an excitement of one of these zones, as for instance on the scalp, hypnotizes them, and an excitement of the same spot awakens them. In this case it may be said that the same cause has produced contrary effects, depending on the physical condition of the subject at the moment of its action.   But this is not a general rule. Some zones are exclusively hypnogenic; others are exclusively the reverse.

If there are numerous ways of producing hypnotism, their efficacy greatly depends upon the conditions.   The first of these is habit.   It has been justly observed that the first attempt to hypnotize a subject nearly always fails, and that it almost invariably succeeds when the experiment has been several times repeated.   It is important to note this fact of hypnotic education. Although absolutely no effect may be obtained at the first *séance*, and the subject may declare that he experienced nothing, yet the attempt has impressed a permanent modification on his nervous system, which will render subsequent attempts more easy.   At first the sleep is tardily produced, then it comes in a few minutes, next in a few moments, and finally almost instantaneously.   After this, the subject is entirely in the magnetizer's power.   It is interesting to observe that these facts are the expression of a general physiological law—the law of repetition.   Numerous psychometrical experiments have shown; first, that when an act is frequently repeated, with sufficient intervals of repose, each series of repetition is accompanied by a shortening of the period of reaction; secondly, that this period becomes shorter in proportion to the increase in the number of repetitions; thirdly, that it is finally reduced to its lowest limit.

We now come to one of the questions most disputed at this time in the history of animal magnetism, namely, whether every individual is capable of hypnotization by the processes of which we have given a general account, or if, in order to effect the result, a morbid predisposition must exist in the subject. Is there, to use Ladame's expression, *an hypnotic neurosis*, without which hypnotization is impossible, and are nervous diseases, and especially hysteria, to be regarded as the indispensable predisposition?

We have already said that, as far as its production is concerned, artificial cannot be separated from natural sleep, and we will add that in its attenuated forms the one does not differ from the other in nature and character. We readily admit that artificial sleep may be produced in any subject by repeating, varying, and sufficiently prolonging the attempts, so as to induce fatigue. Before asserting that this result is impossible, these attempts should be made, and it logically rests with the sceptics to prove a negative. It is, however, certain that most nervous patients, and especially those suffering from hysteria, are distinctly predisposed to the hypnotic sleep, and that it differs from natural sleep by special physical characteristics.

It is precisely the addition of such characteristics which constitutes the most important part of the question, for these physical phenomena serve as the indication of the extremely complex psychical manifestations which accompany them.

Up to this time it has been asserted that physical phenomena, impressing a special character on the sleep, have only been observed in the hypnotism of hysterical patients, described by Charcot and Richer under the name of profound hypnotism. We admit that, in a great majority of cases, sufficient exhaustion to cause sleep may be artificially induced. But the following point remains open to discussion: whether, because it is proved that an individual is artificially put to sleep, it necessarily follows that this is a special, not a natural sleep.

Even if this question should be decided in the affirmative, and it should be established that no one is absolutely refractory to hypnotism, we should feel justified in asserting that hypnotic phenomena consist in a disturbance of the regular functions of the organism. As Barth lately observed, it is possible to give every one a headache, but this does not prove that a headache is a physiological state. We do not therefore accept the opinion of those authors who treat hypnosis as a physiological state, and appear more anxious to separate it from other forms of neurosis, than to connect it with them.*

A second question is immediately connected with the former, namely, whether an individual susceptible to hypnotism can be hypnotized without his consent, and even against his will. Many persons are agitated by the idea that a stranger may influence and dispose of them as if they were mere automata. This is certainly dangerous to human liberty, and it is a danger which increases with the repetition of experiments. When a subject has been frequently hypnotized, he may be unconsciously hypnotized in several ways: first, during his natural sleep, by a slight pressure on the eyes; next, in the case of an hysterical patient, by surprising her when awake by some strong excitement, such as the sound of a gong, an electric spark, or even by a sudden gesture. Some curious anecdotes are told on this subject. An

* Under the name of hypnoscope, Ochorowicz has invented an instrument to show the peculiar sensitiveness of some subjects to the magnet. These subjects appear to be also more easily hypnotized.

hysterical patient became cataleptic on hearing the brass instruments of a military band; another was hypnotized by the barking of a dog; another, who had hypnogenic zones on her legs, fell asleep in the act of putting on her stockings. Even supposing that the subject knows that he is to be hypnotized, and desires to resist, this resistance will often be in vain, in spite of his urgent protestations, and he will soon submit to the authority which the experimenter has acquired over him. Sometimes, however, it has occurred to the subject that he will not sleep, and then the experimenter finds himself opposed by an idea which he is unable to modify;—neither the gong nor the electric light produces any effect, and pressure on the eyes, continued for hours, only brings on an attack of convulsions. If these fixed ideas are artificially developed, they form an almost complete obstacle to all attempts at hypnotization. Of this the patients are aware, and sometimes, when they do not wish to be hypnotized by a given person, they cause their companions to hypnotize and suggest to them. Experimenters sometimes adopt similar expedients; and the caskets and talismans which have been given to patients, with the assurance that no one can hypnotize them while they carry these objects about, must be regarded as simply a mode of suggestion.

With respect to persons who have never been hypnotized, and to the question whether they can successfully resist the forcible attempt to put them to sleep, some authors have said that an individual can prevent any one from hypnotizing him, *if he resists*. The *naïveté* of this assertion reminds us of those philosophers who say, "I am free to do this or that, *if I wish it*." Everything depends on whether the subject can exercise resistance

and use his will.   It must not be supposed that because
moral resistance is a psychical function, it is found to an
equal degree in all men.   On the contrary, it varies with
the individual, just as muscular force varies.   The question does not therefore admit of a simple answer.   In
the case of a person who has never been hypnotized, and
is not very susceptible to hypnotism, his consent, and
even his good will are very necessary for the success of
the operation, and without these he cannot be hypnotized.
But some people are excessively susceptible, and in them
the resistance is necessarily slight.   They may be taken
by surprise when naturally asleep and hypnotized by
pressure on the eyes, and in the waking state they may
be intimidated, taken by surprise, and may even receive
dangerous suggestions without being put to sleep.*   Such
persons should guard themselves carefully, since the
seriousness of the danger cannot be denied.

* In confirmation of this statement, we may cite the well-known story
of a girl hypnotized by a beggar called Castellon. She left her father's
house in order to follow him, although regarding him with terror and
disgust, and remained in his power for four days, during which time he
outraged his unhappy victim several times. (Despine, *Étude scientifique
sur le somnabulisme.* 1880.)

CHAPTER V.

SYMPTOMS OF HYPNOSIS.

THE hypnotic sleep, by whatever processes it may have been effected, is displayed under very different aspects: sometimes it is marked by distinct physical characteristics, and is then designated as *profound hypnotism;* at other times it does not differ from the natural sleep, and it is then termed *slight hypnotism.*

Between natural sleep and the most profound hypnosis it is possible to establish an unbroken chain of intermediate states, which it is somewhat difficult to distinguish from each other. The diversity of symptoms which marks the gradation of hypnotic states accounts for the disputes which are of daily occurrence, and which are far from being exhausted. Each observer, who conscientiously describes the subject before him, believes himself to be in possession of the whole truth, and allows himself to doubt the phenomena which he does not find in this instance. In many cases he even denies their existence, thus contributing to establish an absolute disbelief in those who do not observe for themselves.

Without attempting a critical study of these discrepancies, we believe that they may be ascribed to two chief causes: first, the different states of the patients on whom the experiments are tried; second, the variable nature of the exciting causes of hypnotic phenomena in these patients. If the Salpêtrière school obtained results which do not only agree with each other, but with those obtained by other observers (Tamburini, Seppili, etc.), it is because they took care to define with the utmost accuracy the physical conditions of their subjects, and the nature of their experimental processes. These two points include the whole method summed up by Paul Richer * in the following propositions:—

1. To choose those subjects for experiment whose physiological and pathological conditions are well known to resemble each other.

2. To submit the different experimental conditions to a rigorous law.

3. To proceed from the simple to the compound, from the known to the unknown.

4. To guard carefully against simulation.

5. To be chiefly occupied with simple cases, that is, with those in which the different phenomena appear to be most distinct and isolated from each other.

6. To follow the method of nosologists in classing these different phenomena in natural series, so as to establish several subdivisions in the great group of facts collected under the name of hypnotism.

We shall in our description accord the first place to hysterical hypnosis, which is entitled to serve as an introduction to the general study of hypnotism, not only on account of its historic importance, but on account of its clearly marked divisions, and the intensity of its symptoms. We shall describe separately each of the hypnotic

---

* P. Richer, *Études cliniques sur la grande hystérie, ou, Hystéro-epilepsie*, p. 512, 2nd edition. 1885.

symptoms, beginning with the neuro-muscular phenomena, which are manifested by more objective, and to some extent more palpable, signs than the others. We shall substitute synthesis for analysis, and give an account of the different nervous states designated by Charcot under the names of lethargy, catalepsy, and somnambulism. In order to do this, we must define the nature of these hypnotic states, which have been the subject of so much discussion.

Our study of profound hypnotism will be succeeded by that of its slighter forms; we shall endeavour to classify all these different states, and to connect them with each other, so as to show how the phenomena of hypnotism are allied with those of physiology. We hold that hypnotism should not be considered by itself, nor simply as a matter for curiosity; it is chiefly important as enabling us to study the physiological processes in man, and especially the cerebral functions, and it is adapted to play a considerable part in psychology.

We do not propose, however, like some German writers, to discuss theories on the mechanism of the nervous sleep, since these theories, whether physical, chemical, or physiological, are not founded on solid experience, and appear, at all events at present, to constitute the metaphysics of hypnosis. We shall aim at giving to the ensuing descriptions a purely symptomatic character.

I. Neuro-muscular Hyperexcitability.

Definition—Excitement of the Muscles.—Charcot and his school regard this important phenomenon as the dominant characteristic of lethargy.

The patient in a lethargic state appears to be in the deepest sleep; the eyes are closed, or half-closed, the eyelids quiver, the face is impassible and expressionless. The body is perfectly helpless; the head is thrown back; the limbs hang slackly down, and if they are raised and again dropped, they fall heavily back into the same position.

An examination of the muscles shows, however, that they have acquired the property of contracting under the influence of a direct mechanical excitement, and even, when thus contracted, of forming a contracture, that is, of remaining fixed in the acquired position. To this phenomenon Charcot gives the name of neuro-muscular hyperexcitability.* It may be produced by very simple treatment. For instance, on kneading the muscles on the front of the fore-arm, the limb becomes fixed in a bent position; if the thenar eminence is excited, the thumb turns inward on the palm of the hand. If the muscles of the face are excited, those, for instance, which connect the malar bones with the lips, the latter are raised upwards and outwards. It may be said that all the striated muscles respond to mechanical excitement, without excepting those which do not usually contract under the influence of the will, like the muscles of the pinna of the ear. The abdominal and thoracic muscles form no exception to this rule, so that it is imprudent to perform experiments of this kind on hypnotized patients without an accurate acquaintance with anatomy and physiology. Some unskilful experimenters have produced unpleasant phenomena by

---

* J. P. Charcot and P. Richer, *De l'hyperexcitabilité neuro-musculaire* (*Archives de Neurologie*, 1881–1882).

simply touching the larynx, and by manipulating the diaphragm.

In order to produce a lethargic contracture, a mechanical excitement is usually required, which goes beyond the limits of the skin, and either acts directly on the muscles, on the tendons, or on the nerves.

There are several ways of applying the excitement; in most cases friction, pressure, a shock, and massage are equally successful. The process may be carried out equally well with the hand and with an inert body. The application of a magnet, held at a little distance from a group of muscles, produces the same effect as direct mechanical excitement, but with more energy and diffusion.* Finally, the degree of excitement is important; a slight excitement produces a simple contraction, a stronger one produces a contracture.

Excitement of the facial muscles.—The facial muscles, during the lethargy accompanied by neuro-muscular hyperexcitability, are differently affected from the other muscles of the body. Contraction may be produced by mechanically exciting the nerve which animates them, for instance, the facial nerve as it issues from the parotid, or by exciting the body of the muscle itself; but this contraction does not become permanent contracture. It generally ceases with the pressure, and if the excitement is continued for some time, the effect is exhausted, and the muscle becomes relaxed. In order to fix the contraction of the facial muscles during the lethargy, it occurred to one of the present writers to uncover the subject's eyes at the moment when contraction had been effected. The subject at once became

---

* Tamburini and Seppili, *Rivista di Freniatria*, p. 278, 1881.

cataleptic, and the contraction of the muscle which had been excited was maintained for some time.

It is possible to cause many of the muscles to contract singly, such as the frontalis, the depressor alæ nasi, and the triangularis menti. Several muscles may also be contracted simultaneously, so as to produce what is termed by Duchenne combined, expressive contractions. With the finger, or with a slender stick, rounded at the end, all the electric experiments performed by Duchenne on subjects in the waking state, may be reproduced on the face of a subject in the lethargic state. These studies, carried on with the utmost care by Charcot and Richer, afford an experimental proof of the part taken by each muscle in the expression of the emotions. In fact, with some few exceptions, the muscular action due to hyperexcitability is strictly localized in the muscle which has been directly excited; and the action of this muscle does not induce that of the other muscles which are habitually associated with it, in order to produce an emotional expression. For instance, by pressing the finger, or the end of a blunt pencil on the zygomaticus major, an isolated contraction of this muscle may be effected, so as to give the expression of a forced laugh. In order to obtain the expression of spontaneous laughter, the inferior half of the orbicularis palpebrarum must be simultaneously excited. Lastly, the hyperexcitability of the muscles of the face make it possible to set in motion those muscles which are not usually subject to the will, such as those of the pinna of the ear.

The contraction of the muscles is not only produced by acting on their fleshy body; the mechanical excite-

6

ment of their tendons, or fibrous extremities, produces the same effect.

Excitement of the Tendons.—The effect of exciting the tendons of the knee is particularly marked. If, in the case of a normal individual, the ligamentum patellæ is struck, a contraction of the quadriceps femoris takes place, and this induces a slight shock in the limb, together with an extension of the leg. Hysterical subjects frequently present in their waking state an exaggeration of this tendon reflex. But some fresh symptoms occur in the artificial lethargy : first, a diffusion of the reflex action which is displayed in the shock extending to all the corresponding half of the body ; and next, by a marked tendency to contracture.

Excitement of the Nerves.—The mechanical excitement of the peripheral nerve-trunks is chiefly interesting from the fact that it produces the contracture of all the muscles to which the excited nerve is distributed. Hence it results that the limb subject to experiment assumes a characteristic attitude, which is determined by the special distribution of the branches of the excited nerve to the muscles of that region. It has been said that neuro-muscular hyperexcitability constitutes an anatomical demonstration of the reality of the nervous sleep; it is at any rate certain that this phenomenon cannot be simulated, even by those subjects who are thoroughly acquainted with anatomy.

The nerves of the arm, which are easily accessible to mechanical excitement, are generally chosen to demonstrate this neuro-muscular property of lethargy. The ulnar nerve may be easily reached, in the region of the elbow in the hollow between the olecranon and the internal condyle. If mechanical pressure is exerted by the finger on this point, the subject's hand becomes contractured in the attitude represented in Fig. 1.

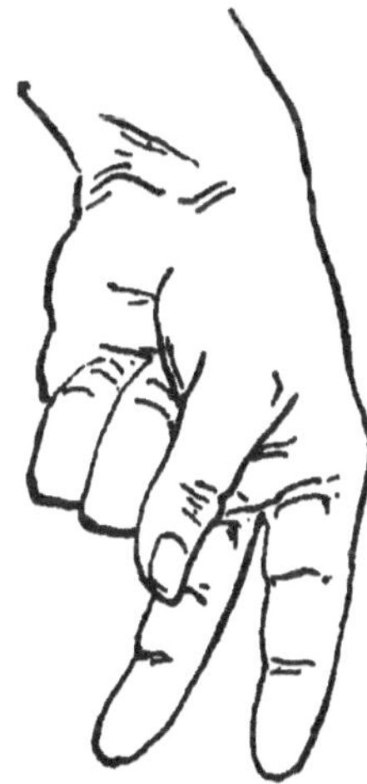

Fig. 1.—Ulnar attitude.   (From Charcot and Richer.)

The fundamental characteristics of this attitude, which presents some secondary variations in different subjects, are the flexion of the ring and little fingers, the adduction of the thumb, the extension and separation of the index and middle fingers. Analysis shows that this attitude is in complete accordance with our anatomical and physiological knowledge. On the one hand, anatomy teaches us the distribution of the ulnar nerve in the fore-arm and the hand ; on the other, physiology shows the partial action of the muscles by means of the ulnar nerve. By combining both these data, we may rigorously infer what attitude the hand ought to take under the combined action of all the muscles brought into play. The attitude deduced by reasoning precisely agrees with the attitude produced during lethargy by excitement of the nerve. The attitude is controlled by the localized faradisation. In healthy individuals faradic excitement of the nervous trunks gives the same results as mechanical excitement in subjects in the lethargic state.

The median attitude, which is produced by exciting the median nerve, which extends along the inner edge of the biceps, consists in a contracture which causes the flexion of all the segments of the limb; the fore-arm is

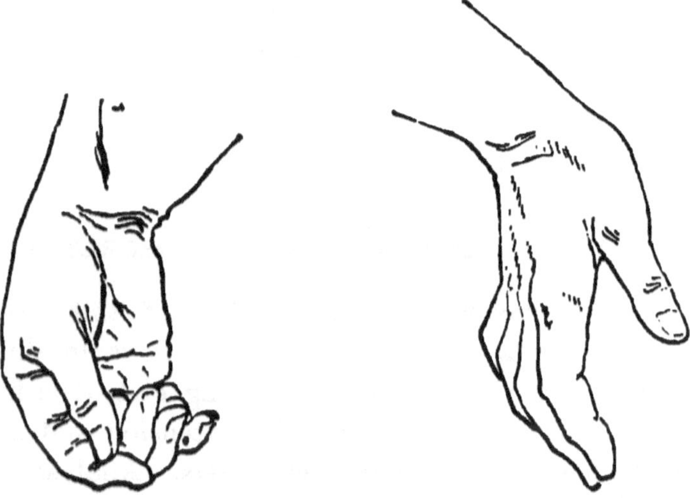

Fig. 2.—Median attitude.   Fig. 3.—Radial attitude.
(From Charcot and Richer.)

raised in a constrained position, the wrist is bent, and the hand closes (Fig. 2).

The radial attitude, which is in some sort the converse of the preceding one, consists in the supine position of the fore-arm, while the wrist and all the fingers are extended. This attitude is produced by exciting the radial nerve, where it issues from the spiral groove of the humerus (Fig. 3).

By the mechanical excitement of the spinal nerves, Berger and Heidenhain were able to produce movements in the limbs in correspondence with them.

We have one remark to make on the localization of the contracture which is produced by exciting the nerve. In the case of the ulnar position, the hand becomes stiffened into what may be termed a sacerdotal attitude. In fact, the muscles in connection with the ulnar nerve are not the only ones affected ; their antagonists also are evidently in a state of tension, and it may be said that all the muscles of the hand are affected. Yet the ulnar attitude assumes a characteristic form which enables us to distinguish it from the median and radial attitudes which we have described. This is due to the fact that, in the collective action of the muscles of the hand, it is only the muscles connected with the ulnar nerve which give a characteristic attitude to the hand, and the other muscles only come into play in order to keep the hand immovable in that attitude ; their contraction is perhaps due to the excitement which affects their fibres in consequence of their sudden extension.

Galvanic Excitement of the Scalp.—The phenomena produced by the galvanic excitement of the scalp of a subject in the state of lethargy must be referred to an hyperexcitability allied to that which is neuro-muscular.

Charcot observed that the application of a galvanic current to the cranial arch during lethargy produced strong muscular shocks in the subject's body. The positive electrode is placed on the scalp, at a level with the motor regions, and the negative electrode is placed on the sternum, on the fore-part of the head, or behind the ear. When the circuit is interrupted, at its opening, and especially at its closure, a distinct shock is produced on the opposite side of the body, and of the face. In some patients the shock occurs on both sides of the body, with a marked predominance on the side to which the electrodes are applied. When the same experiment is performed on subjects in the waking state, variable results are obtained. In some, the galvanic excitement has no effect; in others, its effects are the same as in the lethargic state.[*]

Character of the Lethargic Contracture.—Lethargic contracture presents some characters which clearly distinguish it from a voluntary contraction, and make it easy to ascertain that there is no simulation on the part of the subject.

Experiments have been performed on strong and healthy subjects, who voluntarily assumed attitudes resembling those of lethargic contracture, and the comparison furnished the following results. Under the influence of a continuous traction, the contractured limb of a lethargic subject gradually relaxes, just like the limb which is voluntarily stiffened. So far the resemblance is complete, but the myographic and cardiographic tracings reveal curious differences. In the simulator, the trembling of the limb and the irregular breathing soon betray that the effort is voluntary; in the hypnotized subject the respiratory rhythm does not vary, and the contractured limb is slowly relaxed, without the slightest irregularity.

Charcot and Richer state that when, during lethargy, a group of muscles is excited, and at the same time the limb is not allowed to move in the direction of the

* J. M. Charcot, *Société de Biologie*, January 7 and 14, 1885.

muscles under excitement, this excitement is transferred to the antagonist muscles. For instance, if while exciting the extensors of the fingers the hand is kept half bent, its flexion is accentuated by the contraction of the flexors, although the excitement was limited to the extensors. We remarked above on an analogous fact; the attitude due to a lethargic contracture depends not only upon the muscles which are excited, but also on the antagonistic muscles. It may be stated as a rule of motor-nerve power, that the antagonist shares in the excitement of any muscle whatever. In ordinary circumstances, this contraction of the antagonist has only a regulating function, but it may become preponderant if the effect of the direct contraction is in any way arrested.*

If the contracture is left to itself, it will continue throughout the lethargy; in some subjects the transition to another phase of sleep, or the awakening, will put an end to the contracture; in others, it will remain for an indefinite time, even after they are awake. In order to put an end to it, the experimenter must in this case throw the subject into a fresh lethargy, and then proceed to excite the antagonistic muscles.

Friction and the kneading of the muscles will, in fact, soon relax lethargic contractures. When a contracture of the flexors has been produced, the excitement of the extensors on the back of the hand will soon cause it to disappear. If the sterno-mastoid muscle has-been excited, so as to produce a rotation of the head in the opposite direction, the excitement of the opposite muscle will bring back the head to its original

* Charcot and Richer, *Brain*, October, 1885.

position. This antagonistic action is one of the characteristics peculiar to contractures of the lethargic type.

There is another interesting phenomenon which should not be omitted in the history of neuro-muscular hyperexcitability. Under the name of "a paradoxical contraction," Westphal has described the following phenomenon :—When a sudden and energetic movement of dorsal flexion is, for instance, given to the foot, the anterior tibial muscle contracts so as to produce adduction and a certain degree of dorsal flexure of the foot, which remains fixed in this position. Charcot shows that this phenomenon is more marked in hyperexcitable patients. If, instead of abruptly bending the limb, it is gently placed in the same position, and the extensor muscles are mechanically excited, the limb remains fixed in the attitude of flexion. The excitement of the extensors has a reflex action on the flexors to which they respond by forming a contracture. Erlemeyer makes the reasonable suggestion that the term "contracture by antagonistic distension" should be substituted for that of "paradoxical contraction." This phenomenon, which is most marked in hysterical and hyperexcitable subjects in the state of lethargy, explains why some of these subjects retain the positions due to a sudden effort, as, for instance, when a subject who has thrown a stone, or given a blow, retains his arm in contracture in that position.[*]

The æsthesiogenic action on lethargic contracture must be briefly noticed. In subjects sensitive to the magnet, the transfer of unilateral contractures may be

[*] Ch. Féré, *La Contraction paradoxale* (*Progrès Médical*, 1884, p. 69).

effected by means of this agent; thus, when the ulnar attitude has been produced in the right hand, and a magnet is brought close to the subject's fore-arm when he is asleep, and even when he is awake, both his hands become agitated with slight, jerking movements; then the contracture of the right hand ceases, and is transferred to the left hand, without losing any of its characteristics or of its precise localization. Several other agents, such as a vibrating tuning-fork, metals, and electricity in all its forms, may be used to effect the transfer.[*]

Some interesting phenomena are allied with this last experiment. If the circulation is arrested by the circular compression of a limb in a centripetal direction, by means of one of Esmarch's elastic bandages, the mechanical excitement of the limb thus rendered anæmic does not produce contracture, or rather, it produces a latent contracture, of which there is no external sign, but which is manifested when the circulation returns. In fact, when the bandage is removed, the contracture of the limb takes place in proportion as its colour returns.[†] Again, the magnet applied to the anæmic member transfers the contracture to the sound member, in which it at once becomes visible (Charcot and Richer).

We have observed a phenomenon somewhat allied to the one just cited. When a lethargic subject is placed under the influence of a magnet, and the subject's hand or arm is mechanically excited, the contracture does not occur in the muscle which is directly excited, but in the corresponding muscle of the other arm.

---

[*] K. Vigoureux, *Metalloscope, Metallothérapie Esthésiogènes* (*Archives de Neurologie*, 1881).

[†] Brissaud et Richet, *Progrès Médical*, Nos. 23, 24, 1880.

When the magnet is applied to a bilateral and symmetrical contracture, such as two radial or ulnar attitudes, it does not produce a transfer, but another phenomenon, for which we have suggested the term polarization.* Under the magnetizing influence, both the subject's hands, when in a state of contracture, display slight, irregular, and rapid oscillations, succeeded by more extensive movements, then by actual convulsions, and finally, the two contractures almost simultaneously disappear.

According to Tamburini and Seppili, the neuro-muscular hyperexcitability of a limb may be destroyed by the application of cold water, or of ice.

Neuro-muscular hyperexcitability, like other pathological symptoms, is not equally developed in all subjects. In some we only find an exaggeration of the tendon reflex with no tendency to contracture; in others the contractures may be displayed, yet without any precise localization. Finally,—a singular fact, which shows that in some subjects the waking and hypnotic states are closely allied, and that there are symptoms common to hysteria and hypnosis,—contractures can be easily produced in many hysterical patients in their waking state, either · by kneading the muscles, by pressure on the nerves, or by striking the tendons. These contractures in the waking state are, indeed, of the same nature as those which occur during lethargy, since they yield to the excitement of the antagonistic muscles, and may be transferred by the magnet; they are occasionally as intense and as clearly defined. Several writers—Charcot and Richer, Heidenhain, Tamburini and Seppili, Brissaud

* A. Binet et Ch. Féré, *La Polarisation psychique* (*Revue Philos.*, 1885).

and Richet*—have observed that hyperexcitability may continue during the waking state. In many hysterical patients, digital pressure on the nerves will produce in the waking state median, radial, and ulnar attitudes, identical with those produced in the state of lethargy, with the exception that they are sometimes accompanied by pain. We may infer from these facts, at any rate in the case of some subjects, that an aptitude for contractures is not a symptom peculiar to lethargy, and cannot prove the reality of that state.

In reply to an inquiry into the nature of the contractures produced by muscular hyperexcitability, we should connect them with reflex phenomena, without, however, claiming to throw any vivid light upon the question. Even when the excitement is directly applied to the centre of a muscle, the contracture which ensues is due to a stimulus which has followed the diastallic arc formed by the afferent nerves, the nerve-centres, and the efferent nerves. This is proved by the inhibitory action exerted by the antagonist muscles on the contracture, even when they are, like the sterno-mastoid pair, placed on either side of the median line. This kind of interference can only be produced in the nerve-centres, in the brain, or in the spinal cord. Some of the poisons which affect the central nervous system may, by suspending its action, serve to show the part taken by the nervous centres in neuro-muscular hyperexcitability. If an hypnotized subject is made to inhale ether or chloroform, the moment comes when all traces of hyperexcitability disappear, and the mechanical excitement of the muscles and the motor nerves ceases to take effect.

* *Faits pour servir à l'Histoire des Contractures* (*Progrès Médical,* Nos. 19, 23, 24, 1880).

Finally, neuro-muscular hyperexcitability constitutes the most important objective characteristic of that hypnotic state which is termed lethargy ; it is displayed by an exaggerated reaction to mechanical excitement applied to the muscles, the nerves, and even to the nerve-centres. It cannot, however, be doubted that the same reactions may be produced on some subjects in the state of lethargy under the influence of superficial excitement of the skin, or of bones in the region of muscular insertions. It need not astonish us to find them occasionally in other hypnotic states. We have already observed that neuro-muscular hyperexcitability is displayed in some hysterical patients when not under the influence of hypnotism. In a slight degree, that is, when it is reduced to a simple exaggeration of normal reflex action, neuro-muscular hyperexcitability belongs to other pathological states of the nervous system, with which consequently it is necessary to be acquainted, in order that we may justly estimate the value of this phenomenon.

II. Cataleptic Plasticity.

Immobility is the most striking feature of the cataleptic state. The subject maintains all the attitudes given to his limbs and his body. The arms can be raised or bent by the observer with great ease, since they offer no resistance. The eyes are wide open, the gaze is fixed, and the countenance is expressionless. These collective phenomena give to a cataleptic subject an appearance which cannot be forgotten when once it has been seen.

These attitudes cannot be maintained for an indefinite

time, as some authors have asserted. A cataleptic subject cannot remain in a constrained position for more than ten or fifteen minutes, and a strong man might do as much. The distinctive character of the cataleptic attitudes must be sought elsewhere.

If, in a case of true catalepsy, a tambour is applied to the extended arm to register its slightest oscillations, and a pneumatograph to the chest, to obtain the curve of the respiratory movements (Fig. 4), the following facts may

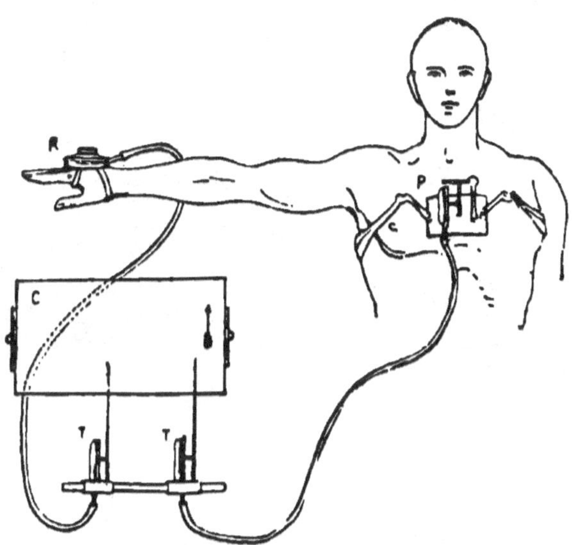

Fig. 4.—Plan of arrangements for experiments in cataleptic immobility. R, Marey's tambour; P, Pneumatograph; C, Revolving cylinder; T T, Tambours with lever. (Charcot, *Leçons sur les maladies du systeme nerveux*, vol. iii.)

be ascertained :—the cataleptic limb does not tremble ; it drops slowly and gently, and the style of Marey's apparatus traces on the cylinder a perfectly regular straight line (Fig. 5, II). At the same time the respiratory tracing maintains the same calm and normal character throughout the experiment (Fig. 5, I). On the

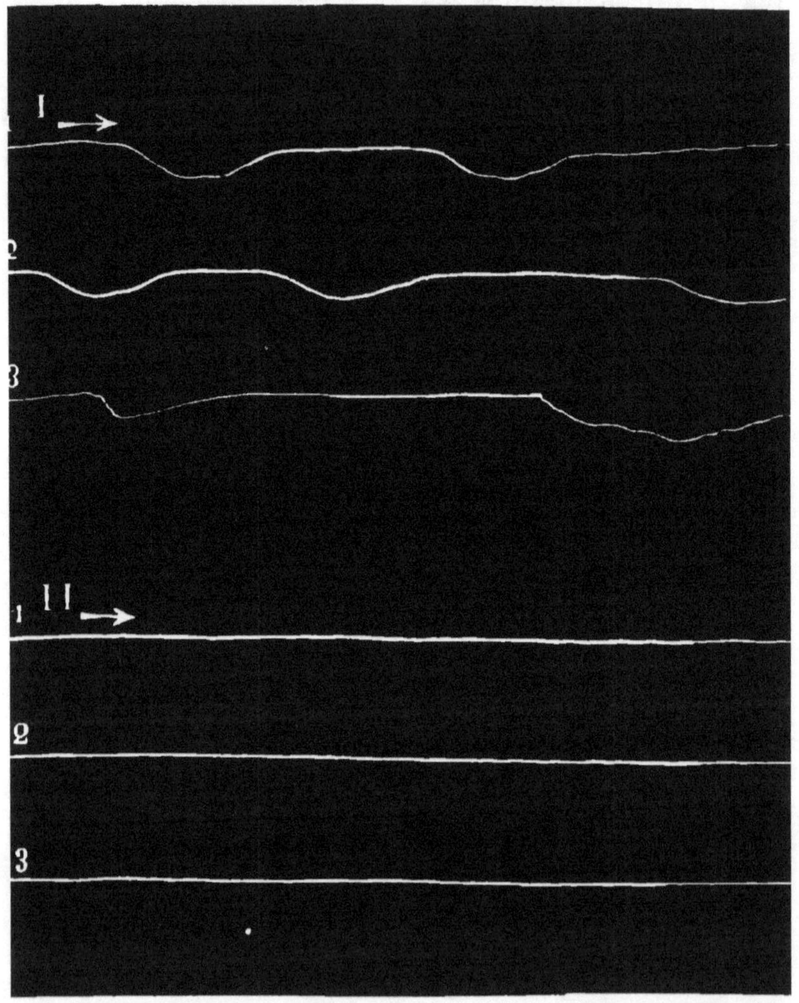

Fig. 5.—Plan of tracings obtained from an hystero-epileptic patient in a state of hypnotic catalepsy (Charcot). I, Tracings of the respiration; II, Tracings of the oscillation of the limb.

other hand, an individual who voluntarily attempts to

maintain such an attitude soon becomes fatigued, his hand trembles (Fig. 6, II); his breathing, calm at

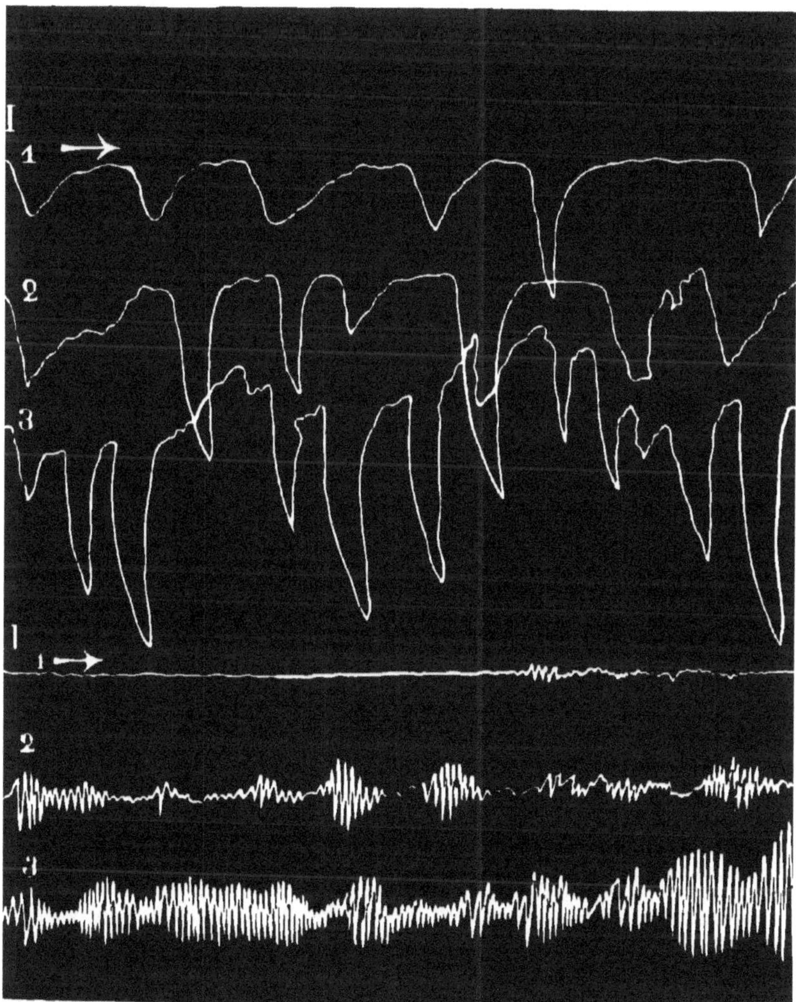

Fig. 6.—Plan of tracings obtained from a man who attempted to maintain the cataleptic attitude (Charcot). I, Tracings of the respiration; II, Tracings of the oscillations of the extended limb.

first, becomes hurried and irregular (Fig. 6, I). The

tracings show abrupt oscillations, which betray muscular fatigue, and the efforts intended to conceal it.

The Salpêtrière experimenters have endeavoured to define the characteristics of true catalepsy, in opposition to the false catalepsy, or catalepsoid states, which may be met with in other phases of hypnotism. If the limb of a patient in a state of lethargy or somnambulism is raised and held up for a few moments, it will remain in the position in which it has been placed. At first sight, this might be called catalepsy, but the truth is that the muscles of the arm were excited by this process, and they have formed a contracture *in situ*. The limb is contractured, not cataleptic; friction and massage will at once cause the muscles to relax. Besides, a certain resistance is offered to a change of attitude, and neither of these characteristics belong to true catalepsy.

We do not, in fact, in profound hypnotism, meet with contractures during catalepsy. If a prolonged pressure is exerted on the muscles, nerves, or tendons, only a relaxation of the muscles takes place, which is followed by paralysis. Richer has devoted himself to the study of cataleptic paralysis. He shows that the paralyzed muscle loses its elasticity and becomes elongated, and the influence of the opposing muscles becomes preponderant. For this reason, when the flexors are excited, the limb is extended. The cataleptic attitude is therefore the exact contrary of the lethargic attitude produced by the excitement of the same motor point. As, however, there is no contracture, the new attitude is not maintained with any rigidity. Localized faradization rapidly puts an end to cataleptic paralysis, if it should continue after sleep is over. It is modified with difficulty by excitement of the antagonists, and by suggestion.

The magnet and other æsthesiogenic agents may effect the transfer of cataleptic attitudes.* A subject is seated near a table on which a magnet is placed; the left elbow rests on the arm of the chair, the fore-arm and the hand are raised in a vertical position, the thumb and fore-finger are extended, and the other fingers are half bent. The right fore-arm and hand are stretched upon the table; the magnet is placed at a distance of about five centimètres, covered by a cloth. At the end of two minutes, the right fore-finger becomes tremulous and is raised, the extended fingers of the left hand become flaccid, and so likewise is the hand for an instant. The right hand and fore-arm are raised and assume the original position of the left hand, which is extended on the arm of the chair with the waxy softness peculiar to the cataleptic state.

It is possible to limit catalepsy to one half of the body, an experiment which it occurred to Descourtes to try at the Salpêtrière in 1878.† It is well known that during catalepsy the eyes are widely opened, and that the cataleptic subject falls into a lethargy if they are closed. If one eye, the right, for example, is closed while the other is kept open, a mixed state ensues; the right side continues to be affected by catalepsy, while the left acquires all the characteristics of lethargy. If the right arm is raised, it retains the position given to it, while the left arm falls heavily down again. Mechanical excitement on the right side fails to produce reflex action, or contracture, while excitement on the left side immediately produces an intense contracture.

Catalepsy may also be combined with somnambulism,

* Ch. Féré and A. Binet, *Société de Biologie,* July 5, 1885.
† *Progrès Médical,* December 21, 1878.

by first throwing the subject into a lethargy, and then acting on one side of the scalp, while opening the eye on the other side.*

The magnet produces the transfer of all these divided states. The transfer of hemi-catalepsy, associated with hemi-lethargy, presents a special feature : at the end of the experiment, the eye remains open on the side which has become lethargic, and conversely, the eye remains closed on the side which has become cataleptic. Thus, in the case of a typical subject, this mode of transfer enables us to obtain a hemi-catalepsy with the eye closed, and a hemi-lethargy with the eye open.†

Cataleptic attitudes display a certain number of characteristics to which we shall revert when we come to describe suggestions. Braid was the first to point out that there is a constant agreement between the attitude of the body and the expression of the countenance. The alternation which exists in catalepsy between the attitudes and the intellectual manifestations should also be noted. When, for instance, a cataleptic subject receives an hallucination, the fixed attitudes, artificially impressed on a limb, give place to complex and perfectly co-ordinated movements, corresponding with the object of the suggestion. The subject resembles a statue, endowed with animation; presently the suggestion is exhausted, the hallucination loses its force, and the subject, if left to himself, again becomes immovable in a cataleptic attitude. This sort of oscillation between psychical and motor disturbance is peculiar to catalepsy.

* Dumontpallier and Magnin, *Société de Biologie*, 1882, p. 147.
† Ch. Féré and A. Binet, *Société de Biologie*, July 5, 1884.

III Cutano-muscular Hyperexcitability.

We have seen that during lethargy strong contractures may be produced by the mechanical excitement of the nerves, of the tendons, or of the bodies of the
muscles themselves, and sometimes also by the excitement
of the skin. In the state of somnambulism, as it is produced in hysterical subjects, we find a contracture which
seems to be of a different kind; it differs both in the
mode of excitement and in the mode of its relaxation.

The starting-point for the contracture of somnambulism appears to be in the skin, which acquires an
exquisite sensibility; it may be produced by making use
of very slight superficial excitements, such as stroking,
passing the hand over the hairs of the skin, breathing
from the mouth, or moving the hand to and fro at a
little distance, so as to induce a slight current of air, and
perhaps also a psychical excitement. This is different
from the contracture of lethargy, which is generally the
result of a strong excitement. This first difference involves a second : produced by a diffused cutaneous excitement, the contracture of somnambulism is itself diffused,
and although it may be limited to one segment of the
limb, there is none of what may be called the anatomical
localization of the contracture of lethargy. On the contrary, the observations of Heidenhain and Dumontpallier
show that it gradually overspreads those parts which
had not been subject to the excitement. But the mode of
relaxation offers the best distinction between these two
species of contracture, at any rate in the typical cases of
profound hypnotism. The excitement of the opposing
muscles, which at once puts an end to the contracture

of hypnotism, has no effect on that of somnambulism ; it can only be relaxed by renewing for a few moments the cutaneous excitement which produced it. Other differences have been noted, but they are less constant than those given above. It has been asserted that it is only the contracture of lethargy which can be transferred by the magnet, but we have been equally successful in the transfer of the contracture of somnambulism.

The aptitude for contracture by means of cutaneous excitement is generally diffused over the whole surface of the body. But it is possible to limit it to a definite region by exciting the scalp in different ways.* We shall presently see, as we continue our description of the different states, that when a subject of profound hypnotism is in a lethargy or catalepsy, friction of the scalp will cause complete somnambulism, and all parts of his body acquire an aptitude for cutaneous contractures. A lateral friction, limited to one side of the head, will produce hemi-somnambulism; restricted to the corresponding side of the body, the state of the other half of the body remains unchanged. Thus we have a hemi-somnambulism, allied with hemi-lethargy, or hemi-catalepsy. If, again, instead of applying the friction to the whole of the scalp, a strong pressure is exerted with the finger, or some blunt instrument, on certain points of the hairy scalp which seem to correspond with the motor centres, it is possible to effect the partial somnambulism of the limb to which the motor centre affected appears to belong. In this way it is possible to effect the isolated somnambulism of one half of the face, one arm, one leg, both arms, both legs, and of

* Ch. Féré and A. Binet, *Société de Biologie*, July 19, 1884.

the whole face. It is even possible to produce the isolated somnambulism of the upper part of the face, by exciting a point of the scalp situated above the horizontal line which would pass through the eyebrows, and behind a vertical line which would pass at the back of the mastoid process, etc. The isolated and successive excitement of these different points produces a generalized, partial state of somnambulism, in which the subject speaks, hears, and is receptive of hallucinations.

The rigorousness of these experiments secures them from fraud, for they involve the local disappearance of the phenomenon of neuro-muscular hyperexcitability which is peculiar to lethargy. This is not a phenomenon capable of imitation; the subject can neither produce nor suppress it at pleasure. We think it is impossible to explain these experiments, and to decide if they are a confirmation of cerebral localization, or if it is to be explained by the existence of reflexogenic zones. The latter interpretation appears to us to be the most probable.

We find, in fact, that in hysterical hypnotized subjects there are several zones in which excitement produces reflex action: first, the hysterogenic zones, on which the pressure produces an attack of hysteria, which is arrested when that pressure is removed; [*] next, the hypnogenic zones, distinct from the former in their position and effects; the excitement of these produces, or in some cases modifies and even puts an end to, the hypnotic sleep. Then come the dynamogenic zones, pointed out for the first time by one of the present writers; [†] the excitement of these produces a momentary

---

[*] Charcot, *Maladies du Système nerveux,* vol. i.

[†] Ch. Féré, *Sensation et Mouvement (Revue Philosophique,* 1886).

exaggeration of muscular force, which may be measured by the dynamometer. There are also erogenic zones, of which we shall speak presently. Finally, Heidenhain, Born, Dumontpallier, and Magnin have described the reflexogenic zones, which, when excited in hypnotic subjects, produce motor phenomena, in places more or less distant from that point on the skin which has been excited. In some of Heidenhain's subjects, pulling the skin of the nape of the neck, in the region of the cervical vertebræ, produced by reflex action a sonorous respiration, or groan; in this way the celebrated experiment performed by Goltz on frogs is repeated on the human subject. Dumontpallier, by exciting the skin of the hairy scalp, produced direct or complex movements, in correspondence with the motor centres excited by him. All these experiments show that in the hypnotized subject many points of the body, and especially those of the hairy scalp, are in a state of hyperexcitability. It would be imprudent to go beyond this simple assertion.

IV. Disturbance of the Breathing and of the Circulation.

When a subject is put to sleep by a slow and prolonged process, as for instance by fixity of gaze, it may be observed that after a while the breathing is quickened; then, at the moment when sleep comes on, a peculiar sound is often heard in the larynx. Tamburini and Seppili have applied the graphic methods of modern physiology to the study of the respiration and the circulation.* The results to which they have arrived by

* *Rivista sperimentale di freniatria*, No. 3, Series vii.; Nos. 3, 4, Series viii.

these methods are in perfect agreement with those made at the Salpêtrière at about the same time.

During the state of lethargy, the respiratory curve is fairly regular; its movements are usually slow and deep; in short, the respiration does not essentially differ from what it is in the normal state. The same may be said of the state of somnambulism. The only characteristic peculiar to hypnotism appears to be a certain disconnection, or even a true antagonism between the thoracic and abdominal respiration.

In catalepsy, however, there is a considerable modifica-

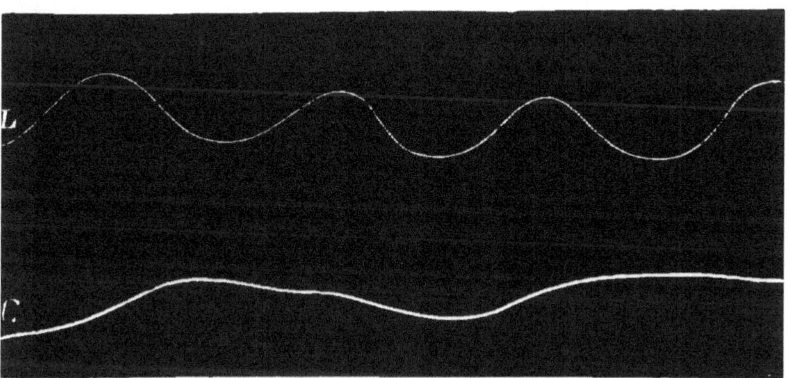

Fig. 7.—Respiratory tracing. L, during lethargy; G, during catalepsy. (Tamburini and Seppili.)

tion in the mode of breathing. The movements are infrequent, superficial, and extremely slow, and separated by a longer or shorter interval of complete immobility. In the subjoined figure (Fig. 7), the widely different tracings afforded by catalepsy and lethargy may be compared.

It has been observed that the application of a magnet to the subject's epigastrum produced profound modifications in the respiratory curve of lethargy; in catalepsy, on the contrary, the curve was scarcely affected by the magnet. The subjoined figure (Fig. 8), which we owe to Tamburini and Seppili, who performed the experiment, accurately represents these two contrary effects. The

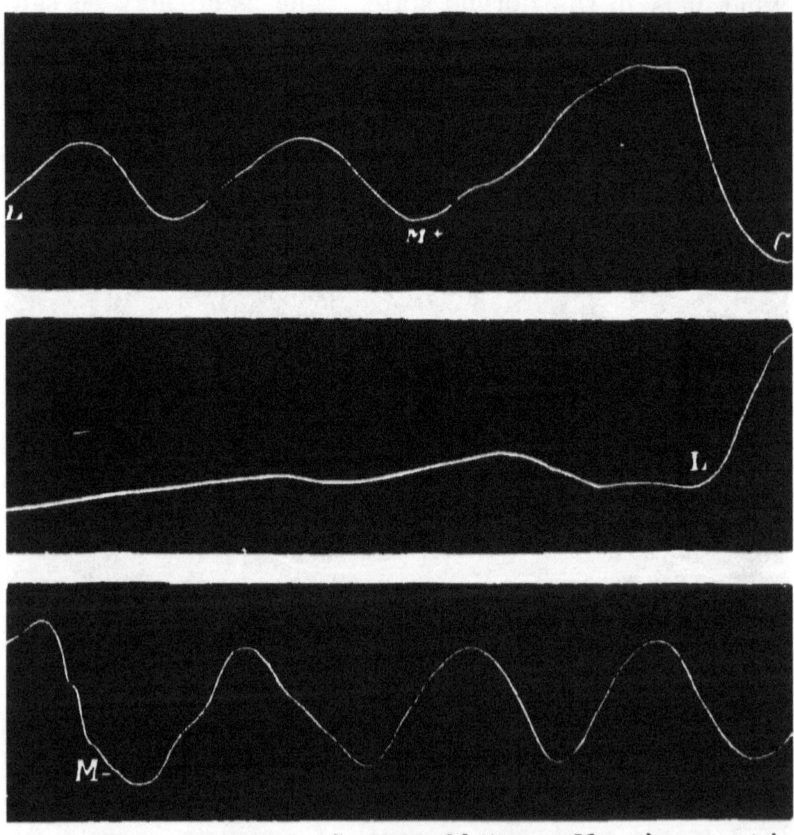

Fig. 8 —Respiratory tracing. L, curve of lethargy; M+, the magnet is approached to the thorax; C, catalepsy is produced; L, lethargy is produced; M —, the magnet is withdrawn.

subject is placed in the state of lethargy; after a few regular respirations the approach of the magnet induces a strong movement of expiration, then of inspiration;

catalepsy is then produced by opening the subject's eyes, and the shallow breathing peculiar to this state is at once displayed. Soon afterwards the eyes are again closed, and lethargy is produced; another deep expiration, followed by a deep inspiration, takes place, owing to the unchanged position of the magnet, and if this is removed, the curve of lethargy reverts to its normal type.

The researches made by Tamburini and Seppili on the circulation are no less interesting. By means of Mosso's plethysmograph, and the air-sphygmograph, they ascertained that in the state of lethargy the graphic tracing shows a constant tendency to rise, and that when catalepsy is produced, it again descends gradually. In other words, lethargy increases the volume of the fore-arm, that is, causes the vessels to dilate; catalepsy, on the other hand, diminishes the volume of the fore-arm, or causes the vessels to contract. Tamburini and Seppili's experiments were repeated by one of the present writers, and although the results obtained were not absolutely corroborative, yet they showed that modifications took place in the peripheral circulation which appeared to be wholly independent of the subject's will.

We have dwelt long upon the neuro-muscular properties of hypnotism, because the Salpêtrière school considers that these phenomena display physical signs which irresistibly prove the sincerity of the experiments. The precise localization of the lethargic contracture in the muscles supplied by the branches of the nerve which has been excited; the maintenance of the cataleptic attitudes without trembling or fatigue; the effects of a continuous traction on the contractures of lethargy and somnambulism; the limitation of each of these phenomena to one half of the body; their mode of appearance and disappearance—all these signs serve as so many guarantees against simulation. On this point the demonstration is complete. It is almost certain that no individual in the waking state, unless affected by a nervous state allied to hypnotism, could imitate the distinctive physical signs by which profound hypnotism is manifested. The dread of simulation, which dominated the whole history of animal magnetism, has now become a completely imaginary danger, if the experimenter is adroit and cautious.

V. Subjective Symptoms.

Up to this time the modifications produced by hypnotism in the condition of the senses, and of the intellectual functions, have not been the subject of accurate research. Some isolated observations have been made, which are not wholly in agreement with each other, and no general view can be deduced from them. In order to obtain a clue to the labyrinth, we should compare hypnotic sleep with natural sleep, and we shall see that the psychical manifestations of hypnosis present a strong analogy to the faculty of dreaming.

1. The state of the senses, in hypnotic subjects, ranges from anæsthesia to hyperæsthesia. During lethargy all the senses are suspended, with the occasional exception of the sense of hearing, which is sometimes retained, as it is in natural sleep. During catalepsy, the special senses are partially awake; the muscular sense, in particular, retains all its activity. Finally, in somnambulism the senses are not merely awake, but quickened

to an extraordinary degree. Subjects feel the cold produced by breathing from the mouth at a distance of several yards (Braid). Weber's compasses, applied to the skin, produce a twofold sensation, with a deviation of 3°, in regions where, during the waking state, it would be necessary to give the instrument a deviation of 18°. (Berger.) The activity of the sense of sight is sometimes so great that the range of sight may be doubled, as well as the sharpness of vision. The sense of smell may be developed so that the subject is able to discover by its aid the fragments of a visiting-card which had been given to him to smell before it was torn up (Taguet). The hearing is so acute that a conversation carried on in the floor below may be overheard (Azam). These are interesting but isolated facts. We are still without any collective work on the subject, of which it would be easy to make a regular study, with the methods of investigation we have at our disposal.

2. More careful observations of the state of the memory have been made, but this state has only been studied as it is found during somnambulism, when it generally displays the same hyperexcitability as the other organs of the senses.

The contrast between the memory on awaking and the memory during hypnotic sleep has been justly remarked. There is a difference between the two phases of memory; and this, indeed, is also the case with natural sleep. The hypnotized subject seldom remembers, on awaking, the events which occurred during his hypnotic sleep. On the other hand, when he is asleep his memory embraces all the facts of his sleep, of his waking state, and of previous hypnotic sleeps.

We will first consider the hyperexcitability of the memory which occurs in somnambulism. Richet performed an experiment which throws a strong light on this strange phenomenon. " After hypnotizing V—— I recited some verses to her, and then awoke her. She was unable to remember them. I hypnotized her again, and she remembered the lines perfectly. When I awoke her, she had again forgotten them."

The memory of a hypnotized subject has a wide range—much wider than it has at other times. Frequent instances of this extraordinary memory have been given, so surprising as sometimes to lead to the belief that the subjects were endowed with a mysterious lucidity. Richet remarks that somnambulists describe with minute details places which they have formerly visited, or facts which they have witnessed. In one instance, a hypnotized subject sang the air of the second act of *l'Africaine*, of which she could not remember a note after she awoke. Beaunis cites the case of a subject whom he induced during sleep to tell him all that she had eaten on the day, or two days before, without omitting a single item. When she awoke, he recounted the *menu* of her dinner, and she was astonished to find him so well informed. We have been able to make a hypnotized subject give the *menus* of dinners she had eaten a week before. Her normal memory did not extend beyond three or four days, and in order to cause her to exceed this limit, it was necessary to use the excitement of the magnet.

We give one more instance, well calculated to display the acute memory of somnambulists. A girl, in a state of somnambulism, was in Charcot's room at the Salpêtrière when Parrot entered, the physician to the refuge for *Enfants assistés*. The subject was asked what was the stranger's name, and she replied, to the surprise of all present, and without hesitation, "M. Parrot." On awaking she declared that she did not know him; but, after looking at him for a long while, she finally said, "I think that he is a physician at the *Enfants assistés*." When about two years old she had been for some time in this refuge, and had long forgotten the physician, whom she now recognized with difficulty in her waking state, while she could, during somnambulism, give his name when ordered to do so.

The acuteness of the memory during somnambulism, without absolutely justifying those who assert that nothing is lost to memory, yet shows that its *conservative* power is much greater than is supposed, when measured by the capacity of reproduction or recollection. It proves that, in many cases in which we believe that a certain fact is completely effaced from the memory, this is by no means the case; the trace of it is there, but the power of recalling it is wanting; and it is probable that under the influence of hypnotism, or of some excitement to which we are sensitive, it would be possible to revive the apparently extinct memory of the fact in question.

It is therefore evident that hypnotism has a peculiar power of exciting the recollection. Our experiments, which are in accordance with those of other observers, tend to show that in the sleeping and waking states the *conservative* memory is about the same. After repeated attempts to make hypnotized subjects repeat a series of figures after only one reading, we could not discover that they were able to retain a greater number of figures than in their waking state. But these are negative experiments, which must not be taken for more than they are worth.

The development of the memory under somnambulism may be compared with its development during natural sleep. There are numerous facts to show that in dreams we see people or hear names with which we were once acquainted, and which we believed we had completely forgotten. Maury, an author who may with advantage be consulted on the subject of sleep and dreams, gives several interesting examples of this revival of old memories in the sleeper. "Some years ago," he writes, "the word 'Mussidan' was recalled to my mind. I knew that it was the name of a town in France, but I had forgotten where it was. A few days later, I saw a person in my dreams who said that he came from Mussidan. I asked him where it was, and he told me that it was in the department of Dordogne." Maury verified the truth of this fact when he awoke. The same author gives another instance of the recall of forgotten facts in a dream. His youth was passed at Trilport, where his father built a bridge. He dreamed one night that he was a child at Trilport, and that he saw a man in uniform and asked his name. The man replied that his name was C——, that he was the bridge-ward, and then disappeared. When he awoke, Maury was haunted by C——'s name, and some time after he asked an old family-servant if she remembered any one of that name. She answered at once that a man named C—— was bridge-ward when his father was at work on the bridge.*

* Maury, *Sommeil et Rêves*, p. 6. Paris, 1861.

The comparison we have just made between natural and artificial sleep may be extended to the phenomena which ensue on awaking. It is well known that the forgetting of dreams is an almost constant fact. At the moment of awakening we have a somewhat vivid sense of our dreams, which is effaced a few instants afterwards, unless we take the precaution of relating them to a third person, or of writing them down. So also in hypnosis; if the sleep has been at all profound, forgetfulness ensues on awaking, and this forgetfulness is even more absolute than after the natural sleep. This characteristic fact has been noted by all observers. Take a subject who has been caused to execute the most complex acts, and to display the most dramatic hallucinations: he has expressed astonishment, has laughed, wept, and been angry —passing through all the violent emotions; he may even have fallen down and injured his head in so doing, yet he remembers none of these things when he awakes. If left to himself, he will be unable to recall one of the scenes in which he has taken part either as witness or actor.

On a closer examination, however, we see that his forgetfulness is not absolute; a vague and confused memory remains, which may be revived by putting the subject on the right road, especially when he is aroused from somnambulism without allowing him to pass through its deepest phases. Heidenhain gives several instances of this recall of the memory, which is, indeed, equally possible in the case of ordinary dreams.

After hynotizing his brother, Heidenhain repeated to him the following quotation from Homer:—

Ποῖον σε ἔπος φύγεν ἑρκὸς ὀδοντῶν.

He then awoke him, and in order to bring the line back

to his brother's mind, it was enough to say, "Homer, flight.' The brother then accurately, but with extreme slowness, repeated the line in question. I take this instance from Richet, who cites another of the same nature : "On awakening F——, I can revive his recollection of what has occurred. He says at first that he remembers nothing, but if, for example, I indicate that he rose up in terror, he says, 'Ah yes, I remember that you made me see a serpent.'" Other experimenters—Beaunis, for example—have made use of a different method, suggestion. It was enough to suggest to some subjects that they would, on awaking, remember all that they had seen, heard, and done during sleep, and their recollection was accordingly complete. Delbœuf arrived at the same result without making any special suggestion; he ascertained that whenever the subject is awakened in the midst of an action, he is capable of remembering all that is connected with that action.* For instance, the experimenter smokes an imaginary cigar beside his hypnotized subject; he suddenly says that the burning ash has fallen from the cigar on to her neckerchief and has set it on fire. The subject rises at once, and places the neckerchief in a basin of water which stands on the table. This is the moment for waking her; she feels that her hands are wet, sees the neckerchief, and recalls the whole scene. In this experiment, the last act of the dream is the first act of awakening. Delbœuf insists on this condition, which he considers necessary to ensure the recollection. It is not enough that the suggestion made during somnambulism should leave a material

---

* Delbœuf, *La Mémoire chez les hypnotisés* (*Revue Philosophique*, May, 1886).

trace; it is also necessary to surprise the subject by awaking him in the midst of an action.

These experiments are the more interesting since they agree with other pathological facts. One of the present writers has shown that in the epileptic state, which has been compared to the so-called unconsciousness of somnambulism, the patient may have retained the memory of the act reputed to be automatic, and can, under the same conditions, even explain it.* We should not, however, be too hasty in including all these modes of reviving the memory in a formula, since the result depends upon many causes—the constitution of the subject, the form of the suggestion, the hypnotic education, etc. It may be a matter of surprise to learn that it is sometimes possible to cause a subject to remember some act committed during somnambulism, without putting him on the right road as Richet does, or giving him a special suggestion like Beaunis, or awakening him in the midst of·an act like Delbœuf; it may be done simply by firmness, and by fixing the subject's attention as steadily as possible on the memory which it is proposed to evoke. If at the same time an exciting cause, such as the magnet, is employed, it contributes to revive the memory by suggestion.

Whatever may be the expedients devised to excite the memory of an individual issuing from the hypnotic state, so as to form a kind of bridge between his sleep and his awaking, the truth remains that a profound hypnotic sleep is always followed by a suspension of the power of memory—a fact proved by the very efforts

* Ch. Féré, *Note pour servir à l'histoire des actes impulsifs des épileptiques* (*Revue de Médecine*, 1885).

which it is necessary to make to restore it. It is evident that hypnosis produces a lesion of the memory.

This lesion is, however, superficial rather than profound; it only affects one portion of the memory—that of the recollection; the memory of conservation remains almost intact, since a fresh sleep gives back to the subject the complete memory which he appeared to have lost in his waking state.

It may, therefore, be said that the disturbance of the memory which ensues from somnambulism is superficial, and only concerns one kind of memory—that of recollection; its power is exaggerated under somnambulism, and depressed on a return to the normal state, and we are still completely ignorant of the cause of such variations. We shall have to make many such confessions of ignorance in the course of this work.

It is difficult to define the intellectual condition of hypnotized subjects; we may estimate the keenness of their senses, and make an inventory of the contents of their memory, but it is not possible to appreciate with the same accuracy the state of their judgment and of their reason. All that can be done is to make the general remark that the intelligence of a hypnotic subject is developed in proportion to his sensitiveness.

What is called lethargy implies a deep and dreamless sleep, in which the psychical faculties are usually dormant. Those subjects who retain the sense of hearing are still capable of receiving some elementary suggestions: if pulled by the sleeve, they may be made to rise, and hallucinations of the hearing may also be produced; but this is all which can be effected. It is, however, possible that lethargy only suspends the power of reaction, and that behind the inert mask of lethargy a remnant of thought is still awake.

In the two other phases of catalepsy and somnambulism, the sleep is not nearly so profound; the subject's intelligence comes into play, and the hypnotic dream begins.

The automatism of catalepsy is its dominant character. This epithet has sometimes been used to define the intellectual character of hypnosis, but it is, in fact, only the cataleptic subject who can be termed an automaton. Catalepsy is sometimes allied with a partial wakefulness of the intelligence, which enables the experimenter to act on his subject by verbal suggestion. In all cases, catalepsy permits the mind to be handled with the same docility as the limbs; the subject's ideality may be said to be plastic. The suggestions offered to him are inevitably accepted, since he never resists them. It has been justly said that a cataleptic subject ceases to have a personality; that there is no *cataleptic ego*. An analogous state may be found in certain dreams to which we surrender ourselves without reflection and without resistance.

The condition of the somnambulist is very different; he is no automaton, but a person endowed with character, aversions, and preferences. For this reason the name of *secondary condition*, in opposition to the waking state, has been given to somnambulism. In this state there is certainly an *ego*. The somnambulist's intellectual condition may be compared to those dreams in which the sleeper actively intervenes, and displays judgment, critical sense, and sometimes even mind and will. There are, indeed, somnambulists who dream spontaneously, and then cease to be *en rapport* with the experimenter.

Setting aside what concerns lethargy and catalepsy, we propose to study some developments of the intellectual state of a somnambulist. Somnambulism is emphatically the medico-legal state, and it is the state in which the aptitude to receive suggestions is the most fully developed.

We have now under observation two subjects, representing the two opposite types of somnambulism—the active and the passive types. The latter remains motionless, with closed eyes, without speech or expression, and, if asked a question, she replies in a low voice. Yet we are confident that this repose of the intelligence is only apparent; the subject retains her consciousness of places and of persons, and hears all that is said in her presence. The other subject is a singular contrast to the one we have just described, since she is in a state of perpetual movement. As soon as she is thrown into a somnambulist condition, she rises from her chair, looks to the right and left, and will even go so far as to address the persons present with familiarity, whether she is acquainted with them or not. On one occasion the photograph of one of these persons was shown to her; she took it, looked for and found the original, and compared him with the photograph, in order to satisfy herself of the resemblance. At another time she spontaneously described some hypnotic experiments which another person had performed upon her a few days before. In short, this subject did not, like the other one, appear to be asleep. These are, however, only appearances, and we must endeavour to examine more closely the psychical state of somnambulism.

In the majority of subjects there is no marked

difference between their normal life and that of somnambulism. None of the intellectual faculties are absent during sleep. It only appears that the *tone* of the psychical life is exaggerated; excessive psychical excitement is nearly always present during somnambulism. This is clearly shown in the emotions. It is, in general, perfectly easy to make a subject shout with laughter, or shed tears. He is deeply moved by a dramatic tale, and even by words in which there is no sense, if they are uttered in a serious tone. It is curious to note the influence of music; the subject expresses in all his attitudes and gestures an emotion in accordance with the character of the piece.

In short, hypnotism does not appear to effect any radical change in the character of those subjects whom we have observed. The intellectual faculties are as active as before. The following is a convincing proof of the exercise of the mind. A patient who had been admitted to the Salpêtrière at an early age was in the habit of *tutoying* M. X—— when she was alone with him, or in company with her acquaintance; she ceased to do so on the entrance of a stranger. Even under somnambulism this patient observed the laws of good breeding, addressing M. X—— as *tu* when she was alone with him, and ceasing to do so as soon as a stranger came in.

It is in somnambulists that we find the curious phenomenon of resistance, of which we shall speak further, when we come to consider suggestions. When an order is given to somnambulists, they will often dispute it, ask the reason, or refuse to obey. It is under the form of a refusal to obey a given order that resistance occurs; subjects more rarely resist hallucinations, since these do not affect their personality. There are, however, instances of this latter form of resistance. When we proposed to transform one of our subjects into a priest, and to give him a cassock, he obstinately refused it. It was suggested to one of Richet's subjects that her arm was being amputated, and she screamed at the sight of the flowing blood, but almost at the same moment she discovered that it was a fiction, and she laughed through her tears. Facts of this kind have unjustly led to the suspicion of imposture. Richet's subject was really under an hallucination, and beheld a sensible image, but her reason was not completely paralyzed, and she was still able to defend herself against the false perception suggested to her.

If we study our own dreams, we may all become aware of these curious duplications of the consciousness; and this shows the connection between normal and hypnotic sleep. The dreamer is, in general, like the somnambulist to whom hallucinations are suggested; he is surprised by nothing, although the most absurd improbabilities are presented to his vision. Yet there is sometimes a remnant of critical sense which induces him to say, in the midst of some grotesque scene, "But this is impossible; I must be dreaming!"

Somnambulists can not only resist, they can tell lies. Pitres states that he suggested to a somnambulist woman that she should murder one of her neighbours, and when she supposed that the crime was accomplished, he caused her, still in the somnambulist state, to appear before a magistrate. She declared her innocence of the crime, and it was only after a prolonged examination, when pressed with questions and overwhelmed by proof, that she finally confessed that she had stabbed her neighbour with a knife. And even then the confession was made with some reserve.*

These facts show that a somnambulist is far from being, as some writers assert, an unconscious automaton, devoid of judgment, reason, and intellectual spontaneity. On the contrary, his memory is perfect, his intelligence is active, and his imagination is highly excited.

Instances have been given of subjects who could, during somnambulism, perform intellectual feats of which they were incapable in the waking state. We ourselves have ascertained nothing decisive on this point, except that we have sometimes observed hypnotized subjects, who could read printing in an inverted position more rapidly than when they were awake, and who could even supply the omitted letters of a double acrostic. There is, indeed, nothing improbable in this quickening of the intellect. There are several instances of a thinker having, when dreaming at night, resolved problems to which he had devoted the fruitless study of many days.

We must, finally, note a peculiar mental state which is only found in slight hypnotism. The subjects assert, on awaking, that they have never for a moment lost consciousness, and that they have in some sense been present as witnesses at the phenomena of suggestion developed by the magnetizer.

The very vague observations to which we have been obliged to restrict ourselves show the difficulty of stating the psychical formula of somnambulism. We are content to assert that the state is not accurately defined by applying to it the term of automatism.

* Pitres, *De la suggestion hypnotique*, p. 63. Bordeaux, 1884.

Finally, their aptitude for suggestions is a feature of the intellectual state of hypnotic subjects, and this fact is so important that we propose to treat of it apart.

The phenomena of elective sensibility, which we have already mentioned in speaking of the processes of hypnotization, are considerably developed during somnambulism. Somnambulist subjects often display a kind of attraction for the experimenter who has hypnotized them by touching the scalp. We shall see presently that friction of the scalp is the means most generally used in the secondary production of somnambulism. When pressure on the scalp is effected with an inert object, as, for instance, with a paper-cutter, a state of *indifferent somnambulism* is generally produced : the subject remains calm, and may be approached and even touched by any one without causing him to make any gesture of defence ; the contractures proper to a state of somnambulism may be produced by any one, or produced by one person and destroyed by another ; they do not depend on any individual influence, and suggestions may be given by any of those present.

It is quite otherwise in the case of *elective somnambulism*. As soon as the experimenter has pressed upon the scalp with his hand, or has breathed upon the subject with his mouth, the latter is attracted towards the experimenter ; if the experimenter withdraws to a distance, the subject displays uneasiness and discomfort ; he sometimes follows the experimenter with a sigh, and can only rest beside him. Any contact with a third person causes suffering.

Elective somnambulism is also produced when the subject is hypnotized by means of passes, which is the practice of magnetizers, or by intimation or suggestion. It is a curious fact that if the subject is told that he will fall asleep at a given hour on the following day, the sleep which occurs at that hour in the operator's absence is *elective*, and the subject is only *en rapport* with the person by whom the suggestion was made (Beaunis). Finally, when the subject is in a state of indifferent somnambulism, and a person touches an exposed part of the body, such as the hands, the elective phenomena are displayed in his favour. All these processes display the common characteristic of bringing the personality of the experimenter into play, and if his importance was formerly exaggerated, it has been too much depreciated since Braid's time.

It has been ascertained that electivity is altogether absent in some subjects, while it is constant in others. And again, in addition to the artificial electivity developed by the experimenter, there is a natural or spontaneous electivity; for this reason one experimenter is more successful than another in hypnotizing or in giving suggestions to a given subject, and especially when he has often hypnotized that subject before.

This special influence of one individual on another, which is so strongly marked during somnambulism, is, in fact, only the exaggeration of a normal fact. It is not uncommon to find persons who feel a special attraction towards some others, and who have a sense of sympathy or antipathy without any sufficient motive. It cannot be disputed that these are real psychical states, although psychologists have for the most part abandoned their study to novel-writers.

It is probable that the phenomena of electivity have

their origin in the experimenter's contact with his subject. Bain, in his work on the Emotions, remarks that animal contact and the pleasure of an embrace are the beginning and end of all the tender emotions. We have seen, in fact, that electivity is displayed in a somnambulist after his scalp and bare hands have been touched by the experimenter; the action of the fingers, as they are used in making passes, seems also to be due to a like influence. The production of elective somnambulism by means of suggestion may also be explained by the fact mentioned above, that since suggestion consists in the recall of a sensation, it probably acts in the same way as a sensorial excitement.

An ingenious experiment made by Richer confirms this view, and shows that electivity has its source in an exaggeration of the sense of touch. "When the subject is in a state of profound somnambulism, owing to friction of the scalp with some inert object, two observers come forward, and each takes hold of one of his hands, without meeting with any resistance on his part. Very soon the subject presses each observer's hand with his own, and will not leave go of them. The special state of attraction applies to both, and the subject is in some sort torn in two. Each observer only possesses the sympathy of one half of the subject, who offers the same resistance to the observer on the left, when he attempts to seize the right hand, as to the observer on the right, who would take the left hand." *

A variation on this experiment is also very instructive. The experimenter preferred by the subject may transmit this attraction to another person; the

* Richer, *op. cit.*, p. 663.

second experimenter has only to slip his hand over that of the former one, so as gradually to lay hold of the subject's hand, and he, after one strong shock, presses up to him in the same way as to the first experimenter.

Elective sensibility is displayed by several phenomena, of which that which relates to contractures is the most important. Only the experimenter who is *en rapport* with the subject can produce and destroy the contractures of somnambulism. It is useless for another person to try to put an end to a contracture by a fresh excitement of the same nature, directed on the same point. His efforts are fruitless, even when the subject cannot see him. The hyperæsthesia of the sense of touch enables the subject to recognize the contact of one operator in a thousand; he may even recognize it through his clothes.

Electivity is also found in suggestions. In the case of indifferent somnambulism, the subject complies with all suggestions, from whomsoever they come; an hallucination effected by the words of one person may be continued by another, and destroyed by a third. This also occurs in catalepsy. In elective somnambulism, the subject is often only able to hear the voice of his hypnotizer, and from him alone he can receive suggestions. We have also remarked that when two observers divide the subject's sympathy in half, the hallucination by the one *en rapport* with the right side only affects the right eye; it is unilateral, and the subject sees nothing with his left eye.

When the phenomena of elective sensibility are subjected to æsthesiogenic action, repulsion, by a singular transformation, succeeds to attraction. At the moment when one of our somnambulist subjects was holding M. X——'s hands, we placed a small magnet close to his head. The subject at once withdrew from M. X——, uttering a cry; M. X—— followed her; she still withdrew, groaning whenever he touched her. Shortly afterwards she came towards the experimenter of her own accord, and again drew back, so that it was impossible to touch her. When she approached for the third time, he took the opportunity of awaking her.*

We must here remind our readers that in the case of some hysterical subjects there are regions in certain parts of the body, termed by Chambard *erogenic zones*,† which have some analogy with the hysterogenous zones, and simple contact with these, when the subject is in a state of somnambulism, produces genital sensations of such intensity as to cause an orgasm. These phenomena have often been displayed, unknown to the observer, who might be liable to the gravest imputations, unless he had taken the precaution, indispensable in such cases, of never being alone with his subject. When we add to this fact the possibility of suggesting to the somnambulist the hallucination that some given person is present, it is easy to see what culpable mystifications might occur.

The erogenic zone only becomes sensitive when somnambulism is absolute. In partial somnambulism, produced by artificial excitement in the region of the motor centres of the limbs, the erogenic zone is inactive; it becomes active when the occipital region of the brain is excited.

* Binet and Féré, *La Polarization psychique* (*Revue Philosophique*, April, 1885).
† Chambard, *Études sur le Somnabulisme provoqué.* 1881.

The erogenic zone may be transferred by the magnet. This transfer is followed by consecutive oscillations, which produce an intense genital agitation. Finally, the excitement of the erogenic zone has no effect unless it is made by a person of the opposite sex; if the pressure is made by another woman, or with an inert object, it merely produces an unpleasant impression.

CHAPTER VI.

THE HYPNOTIC STATES.

Charcot's nosographic essay—Catalepsy—Lethargy—Somnambulism—
Meaning of the three states—Their variations—Intermediate states.

THE different phenomena presented by the symptoms of hypnotism may either exist separately or occur associated in a certain order.   Charcot and his pupils have observed that in hysterical subjects these symptoms tend to fall into three distinct groups.   We think it well to give here a summary of Charcot's nosographic essay.*

"*Attempt to make a nosographic distinction of the different nervous states known under the name of Hypnotism.*

"The numerous and varied phenomena which are observed in hypnotic subjects do not occur in one and the same nervous state.   In reality, hypnotism clinically represents a natural group, including a series of nervous states, differing from each other, and each distinguished by peculiar symptoms.   We ought, therefore, to follow the example of nosographists in endeavouring to make a clear definition of these different nervous states, accord-

* *Comptes rendus de l'Académie des Sciences.* 1882.

ing to their generic characters, before entering on the closer study of the phenomena presented by each of them. It is owing to not having begun by defining the special state of the subject under observation that observers so often misunderstand and contradict one another without sufficient cause.

"These different states which, taken as a whole, include all the symptoms of hypnotism, may be referred to three fundamental types : 1st, the cataleptic state ; 2nd, the lethargic state ; and 3rd, the state of artificial somnambulism. Each of these states, including moreover a certain number of secondary forms, and leaving room for mixed states, may be displayed suddenly, originally, and separately. They may also, in the course of a single observation, and in one subject, be produced in succession, in varying order, at the will of the observer, by the employment of certain methods. In this latter case, the different states mentioned above may be said to represent the phases or periods of a single process.

" Setting aside the variations, the imperfect forms, and the mixed states, we do not propose in this account to do more than indicate briefly the general features of these three fundamental states, which may be said to dominate the complex history of the symptoms of hypnotism.

"1. *The Cataleptic State.*—This may be produced : (*a*) primarily, under the influence of an intense and unexpected noise, of a bright light presented to the gaze, or, again, in some subjects, by the more or less prolonged fixing of the eyes on a given object; (*b*) consecutively to the lethargic state, when the eyes, which up to that moment had been closed, are exposed to the light by raising the eyelids. The subject thus rendered cataleptic is motionless and, as it were, fascinated. The eyes are open, the gaze is fixed, the eyelids do not quiver, the tears soon gather and flow down the cheeks. Often there is anæsthesia of the conjunctiva, and even of the cornea. The limbs and all parts of the body may retain the position in which they are placed for a considerable period, even when the attitude is one which it is difficult to maintain. The limbs appear to be extremely light when raised or displaced, and there is no *flexibilitas cerea*, nor yet what is termed the stiffness of a lay figure. The tendon reflex disappears. Neuro-muscular hyper-excitability is absent. There is complete insensibility to pain, but some senses retain their activity, at any rate in part—the muscular sense, and those of sight and hearing. This continuance of sensorial activity often enables the experimenter to influence the cataleptic subject in various ways, and to develop in him by means of suggestion automatic impulses, and also to produce hallucinations. When this is the case, the fixed attitudes artificially impressed on the limbs, or, in a more general way, on different parts of the body, give place to more or less complex movements, perfectly co-ordinated and in agreement with the nature of the hallucinations and of the impulses which have been produced. If left to himself, the subject soon falls back into the state in which he was placed at the moment when he was influenced by the suggestion.

"2. *The Lethargic State.*—This is displayed: (*a*) primarily, under the influence of a fixed gaze at some object placed within a certain distance of the eyes; (*b*) in succession to the cataleptic state, simply by closing the eyelids, or by leading the subject into a perfectly dark place

"At the moment when he falls into the lethargic state, the subject often emits a peculiar sound from the larynx, and at the same time a little foam gathers on the lips. He then becomes flaccid, as if plunged in deep sleep; there is complete insensibility to pain in the skin, and in the mucous. membrane in proximity with it. The organs of the senses sometimes, however, retain a certain amount of activity; but the various attempts which may be made to affect the subject by means of suggestion or intimidation are generally fruitless. The limbs are relaxed, flaccid, and pendent, and when raised they fall back again as soon as they are left to themselves. The pupils are, on the other hand, contracted, the eyes are closed or half-closed, and an almost incessant quivering of the eyelids may usually be observed. There is an exaggeration of the tendon reflex; neuro-muscular hyperexcitability is always present, although it varies in intensity. It may be general, extending to all the muscles of the animal system, the face, the trunk, and the limbs; and it may also be partial, only present, for instance, in the upper limbs, and not in the face. This phenomenon is displayed when mechanical excitement is applied to a nerve-trunk by means of pressure with a rod or quill; this causes the muscles supplied by this nerve to contract.

"The muscles themselves may be directly excited in the same way; somewhat intense and prolonged excitement of the muscles of the limbs, trunk, and neck produces contracture of the muscles in question; on the face, however, the contractions are transitory, and do not become established in a state of permanent contracture. Contracture may also be produced in the limbs by means

8

of repeated percussion of the tendons. These contractures, whether produced by excitement of the nerves
or muscles, or by percussion of the tendons, are rapidly
relaxed by exciting the antagonist muscles. As it has
been already said, the cataleptic state can be instantaneously developed in a subject plunged in lethargy,
if while in a light room the upper eyelids are raised
so as to expose the eyes.

"3. *The State of Artificial Somnambulism.*— This state
may, in some subjects, be immediately produced by fixity
of gaze, and also in other ways which it is not now
necessary to enumerate. It may be produced at will in
subjects who have first been thrown into a state of
lethargy or catalepsy, by exerting a simple pressure
on the scalp, or by a slight friction. This state seems
to correspond with what has been termed the magnetic
sleep.

"It is difficult to analyze the very complex phenomena which are presented under this form. In the researches made at the Salpêtrière, many of them have been
provisionally set aside. The chief aim has been to define,
as far as possible, the characteristics which distinguish
somnambulism from the lethargic and cataleptic states,
and to demonstrate the relations which exist between
it and the two latter states.

"The eyes are closed or half-closed; the eyelids
generally quiver; when left to himself the subject seems
to be asleep, but even in this case the limbs are not in
such a pronounced state of relaxation as when we have
to do with lethargy. Neuro-muscular hyperexcitability,
as it has been defined above, does not exist; in other
words, excitement of the nerves or of the muscles them-

selves, and percussion of the tendons, do not produce contracture. On the other hand, various methods, among others, passing the hand lightly and repeatedly over the surface of a limb (mesmeric passes), or, again, breathing gently on the skin, cause the limb to become rigid, but in a way which differs from the contracture due to muscular hyperexcitability, since it cannot, like the latter, be relaxed by mechanical excitement of the antagonist muscles; it also differs from cataleptic immobility in the resistance encountered in the region of the joints, when the attempt is made to give a change of attitude to the stiffened limb. To distinguish this state from cataleptic immobility, strictly so called, it is proposed to distinguish the rigidity peculiar to the somnambulist state by the name of *catalepsoid rigidity;* it might also be called *pseudo-cataleptic.*

" The skin is insensible to pain, but this is combined with hyperæsthesia of some forms of cutaneous sensibility, of the muscular sense, and of the special senses of sight, hearing, and smell. It is generally easy, by the employment of commands or suggestion, to induce the subject to perform very complex automatic actions. We may then observe what is strictly called artificial somnambulism.

" In the case of a subject in a state of somnambulism, a slight pressure on the cornea, made by applying the fingers to the eyelids, will change that state into a lethargy accompanied by neuro-muscular hyperexcitability; if, on the other hand, the eyes are kept open in a light room by raising their lids, the cataleptic state is not produced."

We ought to add that this description is made from

nature, and that the Salpêtrière nearly always furnishes patients in whom it is easy to observe these three states, with all their characteristics. In order to observe these states in a new subject, the conditions laid down by the Salpêtrière school must be observed. These two conditions have been already noted by us: (1) The experiment must be tried on the same kind of subject, that is, on one affected by epileptic hysteria; (2) the same mode of operation must be used, that is, by the simplest processes—by fixity of gaze, pressure on the scalp, the electric spark, etc. Any change effected in one of these two conditions alters the experiment and consequently modifies its results.

It must be admitted that even in the case of subjects affected by epileptic hysteria, results differing from those of Charcot will be obtained if the patients are subjected to a different *modus operandi;* if, in other words, they do not receive the same hypnotic education.

We have often been struck by this fact in the course of our researches, and it has appeared the more significant to us, since our experiments have been made on subjects resembling those who served to establish the theory of the three states. We give some examples. It is not, as might be supposed, a necessary symptom of catalepsy that the eyes should be open. We have observed that if hemi-catalepsy and hemi-lethargy are produced, and these hemi-states are then transferred, half of the body becomes cataleptic, although the eye belonging to that half remains shut. Catalepsy with closed eyes may, therefore, exist in profound hypnotism. So, again, it is possible to throw the same subjects into a deep lethargy, in which no trace of neuro-muscular hyperexcitability remains.

We have ascertained that when a magnet is brought near to the arm of a subject in a natural sleep, or to the scalp of a subject in the lethargic state, a new state is produced which has nothing in common with the lethargy described above except the relaxed state of the muscles; mechanical excitement of the nerves, muscles, and tendons, and pressure on the hypnogenic or hysterogenic zones, produce absolutely no effect. No change occurs when the eyes are forced open, the breathing is imperceptible, and there is complete insensibility; it is, in fact, the image of death. Pitres * had the opportunity of observing a case of equally profound lethargy in a patient who was subject to spontaneous attacks of sleep. When one of these attacks came on while he was in a lethargy accompanied by hyperexcitability, this phase of hypnosis became more profound, and all muscular reaction disappeared. Finally, as we have already remarked, neuro-muscular hyperexcitability is not a symptom peculiar to lethargy; in cases of profound hypnotism, contractures may be produced in the waking state, corresponding in all respects to those of lethargy.

These facts only prove that the general symptoms of profound hypnotism may be incomplete or modified, and this is also the case with all other morbid symptoms.

The number of states or periods may also vary in the case of each subject. Speaking generally, there are three states—lethargy, catalepsy, and somnambulism; but this number is not fixed. Dumontpallier and his pupils demonstrated some time ago, and any one may verify the fact for himself, that there are transitional stages between each of these periods, really mixed states, which

* *Des Zones hypnogènes,* p. 65.

the experimenter may make permanent by the employment of appropriate means. In this way from six to nine new states may be created, or even a greater number. It is probable that the invention of new experimental processes, subjecting hypnotic patients to fresh modes of excitement, would lead to the production of entirely new manifestations, differing from those which have been described up to this time. In fact, hypnosis is not a spontaneous neurosis, but an experimental nervous state, of which the symptoms may vary with the processes which give rise to it, while, however, still falling within the limits of the general physiology of the nervous system.

We should misunderstand Charcot's description if we regard it as a systematic work. The only object of the description was to represent hypnosis in all its forms and details. It must not be forgotten that at the time it was made, he wished to establish the real existence of a certain number of hypnotic phenomena, and to demonstrate the existence of an experimental nervous state by such strongly marked characters as to be obvious to every one. Charcot selected subjects in whom these characters were displayed in an exaggerated form which left no room for doubt. This method was perfectly successful, since even those who were unwilling to accept profound hypnotism, were led to study its less developed forms.

The theory of the three states, therefore, only includes one part of the truth, but it is a part which opened the way to all the researches subsequently made upon the question, and even now profound hypnotism is the only state in which we find such objective characters as to limit the field of discussion. It is the object of the Salpêtrière school, not so much to give a definitive description, as to show that hypnotism may be studied in accordance with the most improved processes of clinical science and experimental physiology, and that the science can only be constituted by means of the characters determined by this mode of study. As long as patients affected by acute hysteria exist, most of the results obtained by the Salpêtrière school may be verified.

The history of profound hypnotism serves as an invaluable guide in threading our way through the confused mass of observations which are not included in this form of neurosis.

CHAPTER VII.

IMPERFECT FORMS OF HYPNOSIS.

Imperfect states—Confusion of states in hysterical subjects—Hypnosis in healthy individuals : Experiments by Richet, Bottey, and Brémaud—Different results obtained by the Nancy school.

THERE are many hysterical subjects in whom the division of hypnosis into three states cannot be traced. Many observers have pointed out these exceptions to the rule, which are indeed much more numerous than the normal cases; it is only just to add that this fact was first pointed out by the Salpêtrière school. Richet writes : " The neuro-muscular phenomena of lethargy and of somnambulism are often confounded, while the cataleptic state retains its peculiar characteristics. Sometimes the confusion is still greater, and the neuro-muscular phenomena remain the same, whatever be the phase of hypnotism."

Dumontpallier, Magnin, and Bottey * have insisted on this confusion of states. They ascertained that some hysterical subjects display an aptitude for contracture throughout the periods of hypnosis. They also found that there was often a complete confusion between the

* Magnin, *Effets des excitations périphériques chez les hystéro-epileptiques à l'état de veille et d'hypnotisme* (*Thèse de Paris*, 1881); *Le Magnétisme Animal* (Paris, 1884).

two kinds of contractures distinguished by Charcot; excitement of the skin and profound excitement of the muscles produced the same muscular phenomena in all degrees of hypnosis. This phenomenon may also occur under the slightest excitement, such as the ticking of a watch, the noise of a telephone, the wind of capillary bellows, a drop of ether or of warm water, a ray of light falling directly on the skin, or reflected from a mirror. Finally, all the peripheral excitements capable of producing contracture are also capable of putting an end to it.

Pitres has also described another deviation from the normal type in what he terms the catalepsoid state, when the eyes are closed, which he has observed in some of his hysterical patients.

We pass from hysterical hypnosis to the hypnosis of persons who are, or are assumed to be, in perfect health. We mean by these words persons who display none of the well-known signs of hysteria. Many experimenters have observed persons of both sexes, of all ages and conditions, without taking any note of their pathological antecedents, which involve such minute research, that nothing can be said about them without a careful examination.* Richet, who holds that no one is absolutely insensitive to magnetism, pursued this course as early as 1875. He hypnotizes his subjects by exerting a strong pressure on their thumbs for three or four minutes, and then by making passes in a downward direction over the head, forehead, and shoulders. After a while, this prolonged process produces what Richet terms som-

* Ch. Féré, *La Famille nécropathique* (*Archives de Neurologie*, 1884); *Nervous Troubles, as foreshadowed in the child* (*Brain*, July, 1885); Déjerine, *De l'hérédité dans les maladies du système nerveux*, 1886.

nambulism. This state is capable of presenting three degrees of intensity.*

The first degree, the period of torpor, occurs after passes have been made for a period varying from five to fifteen minutes. The subject begins by being unable to raise his eyes, and these become red and moist: sometimes the muscles show a tendency to contracture under mechanical excitement.

The second degree, or period of excitement, is not attained at once, but after a series of magnetizations; in this state the subject is asleep, yet able to answer questions. During this period, hallucinations may be produced, acts may be suggested, and there is forgetfulness on awaking.

The third degree constitutes the period of stupor; automatism prevails, together with insensibility to pain and the muscular phenomena of contracture and of catalepsy, over which the author passes too rapidly.

Some recent writers, Brémaud and Bottey, have returned to this question of the hypnotism of healthy individuals, and have more accurately defined the features of Richet's clinical description. They state that if some other processes are substituted for the passes, such as pressure on the closed eyes or on the scalp, or the prolonged and fixed gaze on some brilliant object, in short, if the means in use for hysterical hypnosis are employed, it is easy to produce in healthy individuals, not only somnambulism, but also lethargy and catalepsy, and there is no sensible difference between these states and those produced in hysterical subjects.

Brémaud was also able to produce in men presumed

* Richet, *L'Homme et l'Intelligence.* Paris, 1884.

to be perfectly healthy, a new state which he terms *fascination*. It is effected by fixing the eyes on a brilliant point; the subject appears to fall into a sort of stupor, he follows the experimenter, and servilely imitates all his movements, gestures, and words; he is also sensitive to suggestion. On the physical side contractures produced by excitement of the muscles may be observed, and cataleptic plasticity is absent.

Brémaud considers that fascination represents hypnotism in its lowest degree of intensity. This nervous state cannot be effected in women, nor in men who have been the subjects of repeated experiments. In proportion to the increase of impressionability in the subject, he over-leaps this first stage, and passes at once into catalepsy.[*]

These experiments are completely at variance with the results to which Bernheim, Liégeois, and Beaunis arrived. Their experiments, like the foregoing, were made on all kinds of subjects, without distinction of age, sex, and pathological condition. Their observations do not, in fact, amount to much, exclusive of the facts of suggestion.

By whatever process a subject is hypnotized, the moment arrives when his eyes are closed, and his arms fall slackly down. In this state the subject can hear the experimenter. Although motionless and with a countenance as inert as that of a mask, he hears everything, whether he remembers it or not when he awakes. Of this we have a proof in the fact that the one word "Awake!" uttered once, or repeated several times, awakes him, without touching or breathing on his eyes.

[*] Brémaud, *Société de Biologie*, 1883, pp. 537, 635; 1884, p. 169.

In this state the subject can receive all sorts of suggestions. If a limb is raised, and he is told that it cannct be lowered, he passively retains it in the attitude in which it has been placed. If, again, a given movement is given to his limbs, that movement goes on indefinitely until it is arrested. In most subjects there is complete anæsthesia; the skin may be pricked with a pin, and they do not appear to be aware of it. Those who remain sensible to pain may be rendered insensible to it by suggestion.

These observers have not ascertained that the act of opening or closing of the eyes, or that friction of the scalp modifies the phenomena in any way, or that it develops them in subjects in whom they cannot be produced by suggestion alone. They have only ascertained that the degree in which subjects are liable to suggestion varies with the individual. Some only close their eyes, with or without torpor; in others the limbs are relaxed, inert, or incapable of spontaneous movement; others retain the attitudes given them; contracture by suggestion and other suggested and automatic movements are displayed in other cases. Finally, automatic obedience, anæsthesia, illusions, and hallucinations mark the successive stages to which suggestion may be carried, of which somnambulism is the culminating point. It is only in this latter state, in which the phenomena of suggestion are most fully developed, that there is forgetfulness on awaking. About one hypnotic subject out of six attains to this degree of profound somnambulism.

It appears that the performers of these experiments at Nancy only observed in their subjects the phenomena of suggestion which belong to somnambulism; they hold that all hypnotism is summed up in suggestion. Not content with explaining what they have observed, they repeatedly manifest the intention of including within the sphere of suggestion the lethargy and catalepsy described by other writers. But we may set aside the interpretation, and content ourselves with the facts. It is very strange that the observers at Nancy have not seen contracture produced in hypnotized subjects by excitement of the nerves, tendons, or muscles. Richet has often met with this common phenomenon in his experiments on healthy subjects; it has been constantly observed by Bottey; Braid himself repeatedly mentions it; and yet Bernheim, whose experiments were performed on similar subjects, is unacquainted with it. If it is true that none of his subjects, whatever be the excitements to which they are subjected, displayed any physical characteristics of hypnosis, and that everything was summed up in the phenomena of suggestion, we are compelled to infer that there is no scientific proof that his subjects were really hypnotized. Our disbelief is not absolute; we do not assert that these observers only had to do with impostors, nor do we throw doubt on their experiments in general, but if we had to make a medico-legal examination of one of their subjects, we should find it hard to decide whether he was truthful or a deceiver.

We are not disposed to believe that subjects at Nancy differ from those at Paris. In reality, the differences are not due to the subjects, but to the experimenters; they come from the mode of culture, and still more from the processes of study. As we have repeatedly said, the results of experiments depend on the methods by which they are carried on. If suggestion is employed as the sole process, only the effects of suggestion will be obtained; and thus it was at Nancy. But if we apply ourselves to the study of physical characteristics, they may sometimes be observed at the outset, and they may also be gradually developed in some other subjects.

CHAPTER VIII.

GENERAL STUDY OF SUGGESTION.

I. Definition of suggestion—Suggestion and dreams—Distinction between the experimental processes by ideas and by peripheral excitement—The persons capable of receiving suggestions—The conditions of receptiveness : mental inertia, psychical hyperexcitability—Suggestion during hypnotism—Suggestion during the waking state—Different kinds of suggestion—Speech—Gesture—Muscular sense—Auto-suggestion—Relation between the idea suggested and the peripheral excitement—Error made by those who see suggestion in everything.

II. The method—Simulation—Voluntary suggestions—Unconscious suggestion—Comparison between the phenomena of suggestion and the facts of positive science.

III. Effects of suggestion—Modifications of vegetative functions—The suggestion of psychical phenomena—Classification—Positive suggestions hallucination, and action—Psychological analysis of their production—Law of negative suggestions—Law of psychical inhibition—Post-hypnotic hallucinations—Forgetfulness of the act of suggestion.

I.

Definition of Suggestion.—We have seen that the hypnotic sleep approximates to natural sleep in its mode of production and in some of its symptoms. This comparison may serve as an introduction to the theory of the facts of suggestion. These facts are at first sight astonishing, unintelligible, and sometimes indeed they appear incredible. The question arises how it should be

possible for one person to exert over another the power
of making him speak, act, think, and feel as it pleases the
experimenter to dictate.

In order to make a systematic study, we must proceed
from what is better known, to what is less known. As
we have repeatedly said, the psychical phenomena of
hypnosis can only be understood when they are compared
with the dreams of natural sleep. The effects produced
on hypnotized subjects by suggestion are nothing but a
dream, produced and directed by the experimenter. This is
a legitimate comparison, since it is possible to modify the
dreams of a person who is naturally asleep. Maury performed some striking experiments on himself to illustrate
this fact.*　He begged a person to remain beside him in
the evening, and as soon as he fell asleep to excite certain
sensations in him, without telling him what they were
to be, and to wake him after giving him time to dream.
These dreams, produced by sensorial excitements, did not
differ from those obtained from hysterical hypnotized
subjects, by means of suggestion. On one occasion
eau-de-cologne was given to him to smell; he dreamed
that he was in a perfumer's shop, then the idea of the
perfume aroused that of the East, and he dreamed that
he was in Jean Farina's shop at Cairo. The nape of his
neck was gently pinched, and he dreamed that a blister
was applied to it, which recalled to mind the physician
who had attended him in childhood. When a hot iron
was brought near his face he dreamed of stokers. When
he was asleep on another occasion, a person present
ordered him in a loud voice to take a match, and he
dreamed that he went voluntarily to find one.

* Maury, *Sommeil et Rêves*, p. 127.

Another resemblance of a different kind may be established between the psychical phenomena of natural sleep and those of hypnotic sleep. As one of the present writers has shown,[*] in the case of many patients the pathogenic idea, the first manifestation of delirium, may originate in the waking state, but it is generally confirmed by the dreams of natural sleep, in which it is re-echoed with added strength. Clinical observation therefore shows that the experiments so easily performed during the artificial sleep may be spontaneously realized in the normal sleep.

The region of suggestion has a wide range. There is not a single fact of our mental life which may not be artificially reproduced and magnified by this means. It is easy to understand how much psychologists may gain from this method, which introduces experiment into psychology.

Before going further, we must define the extent and limits of suggestion, by a more precise definition than the somewhat vague and summary one given above. Strictly speaking, suggestion is an operation producing a given effect on a subject by acting on his intelligence. Every suggestion essentially consists in acting on a person by means of an idea; every effect suggested is the result of a phenomenon of ideation, but it must be added that the idea is an epi-phenomenon; taken by itself, it is only the indicative sign of a certain physiological process, solely capable of producing a material effect.

This characteristic will generally enable us to recog-

* Ch. Féré, *La Medicine d'imagination* (*Progrès Médical*, 1881, p. 309 ; 1886, pp. 717, 741, etc.).

nize what is and what is not suggestion, although this is often a very difficult question. Thus, when striking the tendons, or kneading the muscles makes it possible to form a contracture of the arm of an hysterico-lethargic patient, no suggestion is made, since the contracture results from a physical action, into which the subject's mind does not appear to enter. If, on the other hand, the experimenter approaches the subject, and says, without touching him, "Your arm is bent and stiffened, and you cannot straighten it," the contracture which occurs in consequence of these words results from a psychical action. The experimenter's command only produces its effect by traversing the subject's intelligence; it is the *idea* of contracture, entering into the subject's mind, which has produced the contracture, and this is really suggestion. From this point of view, it may be said that the theory of suggestion infuses new life into the old philosophic question of the action of the mind upon the body, and that at the same time it throws fresh light on the large group, still so obscure, of diseases of the imagination.

We give another example. Like contracture, paralytic movements may be produced in two different ways. If in a hysterical subject the fixed end of a vibrating tuning-fork is applied to certain points of the vault of the cranium, a transitory excitement of motor force takes place in the subject's arm which soon passes into complete and flaccid paralysis.* In this case the paralysis is the direct result of the vibratory movement transmitted by the tuning-fork through the thickness of the skull to the brain: the

* Ch. Féré, *Inhibition et Épuisement* (*Bull. Soc. de Biologie*, 1886, pp. 178, 195, 220).

subject's intelligence takes no part in it; the experiment, although made upon his body, has not affected his mind; there is no suggestion. On the other hand, if the idea is impressed upon the subject that his arm is affected by paralysis, the paralysis which ensues is of a psychical nature, since it results solely from the subject's conviction that he is paralysed. It does not result from a physical shock, or traumatism, but from a phenomenon of ideation. A suggestion has been made.

The analysis of this last example makes it possible to avoid a confusion made by some writers. It has been too readily admitted that every hypnotic process which has its seat in the brain, proceeds from a phenomenon of suggestion, and this has led to the opinion that lethargy, catalepsy, and hysterical somnambulism, which are perhaps due to reflex cerebral action, are the pure and simple products of suggestion. The fact just mentioned disposes of this error. The paralysis induced by physical vibration and the paralysis induced by suggestion probably both result from modifications of the cortical substance of the brain; they are consequently reflex cerebral acts. But the two cases are widely different. Paralysis by suggestion demands the aid of the subject's intelligence, and if the function of ideation were for any reason suspended, this kind of paralysis could no longer be produced.

The study of hypnosis may be divided into two parts, distinguished by the different processes which are employed. The first part includes the hypnotic phenomena produced by physical excitements, or sensations, which were the subject of the two preceding chapters; the second part includes the hypnotic phenomena produced by ideas, that is, the theory of suggestions. These two modes of experimentation are parallel, and it is difficult to say which of the two has the widest extent.

Those liable to Suggestion.—Suggestion does not act with equal intensity on all individuals. If you assure a person who is awake, in normal health, and perfectly self-possessed, that she is hungry, she will reply that you are mistaken; if you try to suggest a visual hallucination by asserting that she has a book in her hands, shew ill declare that she does not see it. The assertion only produces in her mind a slight effect, which is quickly effaced. It produces an idea of the phenomenon, not the phenomenon itself. It is, in short, evident that the suggestion influences a sound person no more than the opening of her eyes would produce catalepsy.

In order that the suggestion should succeed, the subject must be either spontaneously or artificially in a morbid state of receptivity; but it is difficult to define with precision the conditions under which suggestion is possible. Two have been given, of which the first is the mental inertia of the subject. It has been said that in an hypnotic subject the field of consciousness is completely vacant. A state is produced, and since there is no obstacle—neither the power of arrest nor that of antagonism,—the idea suggested dominates the sleeping consciousness. This explanation has been given by Heidenhain, Richet, Ribot, and others, yet we doubt whether it is altogether in harmony with the facts. If the limitation to a single idea is realized in cataleptic subjects, it is much more rare in the case of somnam-bulists. We believe that the aptitude for suggestions is caused by a second phenomenon—by psychical hyperexcitability. In our opinion, if the idea suggested exerts an absolute power over the intelligence, the senses, and the movements of the hypnotized subject, it is especially due to its intensity. But we admit that it is difficult to resolve the question, and we prefer to leave it open.

The number of persons liable to suggestions is immense; the liability is evinced, not only in cases of hypnosis and of natural sleep, but also in some forms of intoxication by alcohol, and haschich, and in the waking state. We only propose, however, to consider hypnotized subjects.

The aptitude to receive suggestions is strongly developed by hypnotism, but, as we have already said, it does not occur in all phases of hypnosis, only in catalepsy and somnambulism. The suggestions made to a cataleptic subject are simple, automatic, inevitable; the reason takes part in those of somnambulism; the subject discusses and enlarges on them, and sometimes even offers resistance. We shall have to consider these shades of differences whenever we have to do with interesting suggestions, and we shall for the most part content ourselves with describing the suggestions of somnambulism.

After awaking, the subject is still sensitive to suggestion; this fact has long been known, and is mentioned by Braid among other writers. Of late years it has been studied by Richet, Bernheim, Bottey, etc. It is possible, not only to make suggestions to subjects in the waking state, but also to persons who have not been hypnotized at all. Learned men have been agitated by these latter experiments, which have aroused in them doubt and dissatisfaction.   They have no difficulty in admitting that suggestions may be made to hypnotized subjects, since they are not in normal health, but they cannot understand how they should be made to individuals who are awake, not under hypnotism, and that this should be done by modes of action in daily use in our relations to one another.   The question arises whether individuals capable of receiving suggestions in their waking state are in common life liable to submit automatically to the influence of others; whether they are weak in mind; what is their physical and moral state, and if there is anything peculiar in their waking state and hereditary antecedents.   These are the questions stated by Janet, and they have not as yet received any reply.   The possibility of making suggestions to normal subjects must, however, be admitted if, as one of the present writers has done, we refer suggestion to the act of attention.   When attention is sufficiently intense, the period of reaction may disappear, and may even become negative; that is, the reaction may precede the excitement.   An intense mental representation, whether arising spontaneously, or induced by suggestion, may therefore produce a reaction irrespective of any excitement.[*]

Different kinds of Suggestions.—If it is the characteristic of suggestion to address itself to the subject's intelligence, it follows that there are as many forms of suggestion as there are modes of entering into relations with another person.

The experimenter may begin by employing spoken or written suggestion.   This is the simplest and most convenient means.   In order to produce an hallucination,

[*] Ch. Féré, *Progrès Médical*, p. 741. 1886.

it is enough to name the imaginary object, to say to the subject, "There is a serpent at your feet!" and the hallucination immediately occurs. Verbal assertion is the process most in use, since in this way everything which human speech is capable of expressing may be suggested, and it is also the most precise.

Gestures, which are often employed by some experimenters, are a very inferior means. They are undoubtedly fairly successful in the case of subjects who have been long under treatment. Without uttering a word, the hallucination of a serpent may be produced in such subjects by making an undulatory movement with the finger, or still more simply, by directing their eyes downwards. It also becomes possible to give orders by means of gestures, to constrain the subject to walk, to follow or approach the experimenter, to make him kneel down, etc. On pointing to a hat, the subject takes hold of it, and on pointing again to his head he puts it on. He may also be made to take something out of the pocket of another person. Pitres mentions the amazing quickness with which he has observed some subjects divine the meaning of the slightest movement of the fingers, lips, or eyes. But these processes are lacking in precision. Although it is probable that it is the psychical and expressive character of the gesture which generally acts upon the subject, that is, that it arouses ideas, we do not certainly know that no other cause is at work. The same must be said of passes, for when suggestions are made by their means, we do not know how it is done. But if suggestion by means of gestures is often vague, it may be very intense. When a verbal suggestion of movement is given to a subject, the image of such movement is aroused in his mind; this image, however intense it may be, must always be less intense than the sensation given to the same subject when the movement in question is executed before his eyes. The result of the two experiments is therefore very different.* It has been repeatedly ascertained that if a dynamometer is placed in a subject's right hand, and he is ordered to hold it with all his might, this verbal suggestion only augments his normal dynamometric force by a few degrees ; but if the action of firmly clenching his fist is imitated before him, his muscular force is not merely increased, but doubled; thus showing that in certain cases the suggestions given by gestures afford more intense results than it is possible to obtain by words only.

It is sometimes useful to combine suggestive gestures with verbal suggestion, or with the presentation of an object. For instance, a real object is presented to the subject, of which the nature is said to be different; he is caused to eat paper, while told that it is a cake; or it is suggested to him that one of the persons present has a false nose. Combining the word with the gesture gives definiteness to the suggestion.

Suggestive gestures address themselves to the sight. The other senses may also receive impressions. If a gong is gently sounded close to the ear of an hypnotized subject, he thinks that he hears bells; and if he is pricked or pinched, the image of stinging creatures may be aroused. But all these processes are inferior to that of speech.

In all cases in which the idea awakened in the subject emanates from the experimenter's direct sugges-

* See Ch. Féré, *Sensation et Mouvement (Études de psycho-méchanique).*

tion, the subject is in a state of direct subjection to him. This state is not opposed to that which we may observe in the waking state; there is only an exaggeration of phenomena, which makes it easier to understand what occurs in subjects held to be of sound mind, who are unconsciously influenced by the will of another person substituted for their own. Indeed we have only to glance at social relations in order to see that individuals fall into two categories—the leaders and the led—that is, the givers and the recipients of suggestions.

It is characteristic of suggestion by means of the muscular sense, that it may be said to have its origin in the hypnotized subject. If his limbs are placed in a tragic attitude, a corresponding emotion is displayed; if his fists are clenched, he frowns with an expression of anger; if his limbs are disposed so as to begin any action, it is carried on by the subject, and in this way he may be made to climb or go on all fours; or if a pen or piece of work is put into his hand, he will write or sew. If his hand is raised, and the fore-finger is bent, the idea occurs to him that a bird is perching on his finger, and this hallucination is developed. A slight movement by the experimenter will carry on this silent suggestion; the subject then imagines that the bird has taken flight, and he runs about the room trying to catch it. If the hands of a female subject are crossed upon her bosom, it suggests the idea that she is holding an infant. All these facts are included in the same formula; the attitude given to the subject's limbs is accompanied by definite muscular impressions, which arouse corresponding ideas in the brain.

Since every suggestion has its origin in a sensorial
9

impression which is experienced by the subject, it is easy to see that this impression may be produced by an external object without the intervention of the experimenter. This mode of suggestion is inconvenient, and therefore rarely employed, but it is sometimes spontaneously displayed. Bennett mentions the case of a butcher who wished to place a heavy piece of meat on a hook above his head; he slipped, the hook caught him by the arm, and he remained suspended. He was taken down half-dead, his sleeve was cut open, and, although he complained of great suffering, as soon as the arm was exposed, it was found to be absolutely intact; the hook had only penetrated his coat-sleeve, This is an instance of suggestion without an experimenter, and many others might be given. We shall speak presently of the paralysis induced in hypnotic subjects by suggestion. It is probable that some cases of hysterical paralysis, termed traumatic,—that is, caused by a shock,—are also due to suggestion, since the patient cannot divest himself of the idea that such a severe shock must produce paralysis.

These latter facts pass gradually into those to which the name of auto-suggestion has been given. There are cases in which suggestion has its origin in the subject's intelligence; in which the suggestion is made by himself. Instead of being the result of an external impression, as in the case of verbal suggestion, the suggestion results from an internal impression, such as a fixed idea, or delirious conception. A few examples will explain this better than dry definitions. A subject imagined that she was opposing by force the hallucination suggested by one of the present writers, and that she had given him a blow on the face. When her supposed adversary entered the room on the following day, she imagined that she saw a bruise upon his check. This hallucination was derived from the former one, just as a conclusion is derived from its premisses, and it may be taken as a type of auto-suggestion. The subject must have unconsciously argued after this fashion: I gave him a blow on the cheek, of which therefore he must bear the mark. Another subject, coming out of a state of profound lethargy, which had only lasted for five or six minutes, imagined that she had been asleep for several hours. We encouraged the illusion by saying that it was two o'clock in the afternoon, although it was in reality nine in the morning. When she heard this, the patient felt extremely hungry, and begged us to let her go to get food. This was a kind of organic hallucination—the hallucination of hunger—suggested to the subject by herself. She unconsciously reasoned somewhat after this manner: It is two o'clock in the afternoon; I have eaten nothing since I got up, and am therefore dying of hunger. This imaginary hunger was soon satisfied by an equally imaginary meal. We suggested that there was a plate of cakes on a corner of the table, of which the subject might partake, and at the end of five minutes her hunger was appeased. These examples of auto-suggestion are derived from hallucination, and we now give one belonging to a different order of ideas. We approached a hypnotized subject, and addressed her as follows: "A serious accident has just befallen you. Do you remember it? Your foot slipped in crossing the court-yard, and you fell upon your hip. You must have hurt yourself very much." The subject instantly felt a severe pain in the hip and began to moan, and also, suggesting to herself the natural consequences of the fall, she gave herself a slight paralysis of the limb, and limped on awaking.

The general conclusion to be derived from all these facts and experiments is, that suggestion consists in introducing, cultivating, and confirming an idea in the mind of the subject of experiment. In reply to the inquiry, What is meant by an idea, and what latent force does it possess in order to affect some individuals so powerfully? we must repeat that the idea resolves itself into an image, and the image into a revival of the sensation. It consists in the psychical renewal of a peripheral sensation already experienced by the subject. This enables us to understand its power; the idea is, strictly speaking, only an appearance, but there lurks behind it the energy excited by a physical, anterior excitement.

This point of view is confirmed by the fact that it is possible to produce, by simple physical excitements, that is, by sensations, almost all the effects which have hitherto been produced by suggestion, that is, by ideas. Thus, instead of producing or putting an end to paralysis by speech, it may be produced by a shock to the limb or on the skull (Charcot), and it may be terminated by employing the same process. The repetition of the vibrations of the tuning-fork represents the shock of traumatism. Hysterical anæsthesia may also be produced and destroyed by analogous processes.* The movements may also in some cases be produced by excitement of the scalp. We cannot here give full details, and only wish to show that suggestion may be referred to the

* Ch. Féré, *Note pour servir à l'histoire de l'amblyopie hystérique* (*Bulletin de la Soc. de Biologie*, 1886, p. 389).

peripheral excitement in which it has its source. It is precisely because it produces in the nervous centres of some subjects the same dynamic modifications, that suggestion is able to effect all the phenomena which result from peripheral excitements. We are reminded of an old saying which is not yet obsolete, *Nihil est in intellectu quod non prius fuerit in sensu.*

In our day the power of suggestion has been so firmly established that some people have maintained that it is to this we must ascribe the action of æsthesiogenic and dynamogenic agents, etc., employed for peripheral excitements. The reality of the process of cure by these agents has been denied, and it is thought that suggestion will explain the singular phenomenon of transfer, discovered by Gellé, and afterwards studied by a commission of the *Société de Biologie;* what Carpenter ascribed to *expectant attention* is now ascribed to suggestion. This error is chiefly due to the idea that if suggestion will reproduce any given phenomenon which was previously ascribed to a physical excitement, therefore suggestion is its true cause. But this argument is weak. It might as well be said that since it is possible to satisfy a somnambulist's hunger with an imaginary meal, nourishment is at all times unnecessary for him.

Moreover, this opinion does not possess the merit of simplicity which is claimed for it, since it is as difficult to understand why the simple idea of paralysis should paralyse,. as to understand why a shock to the skull should produce the same effect. Besides, in ascribing everything to the idea, we ignore the fact that it is a secondary and derivative phenomenon. To maintain that the idea is everything, and the peripheral excitement nothing, would be equivalent to maintaining that the idea is a phenomenon entirely independent of the sensorial functions; it is, in fact, a revolt against the grand theory of the relations between sensations and images which dominates modern psychology. Such an opinion is also opposed to physiology, which teaches us that several functions—the secretions of sweat, tears, etc.—may become active by means both of physical and mental causes, and that the reality of the one does not exclude the other.

II.

We now come to the important question of the method.

Such a book as this cannot include within its narrow limits the innumerable details of hypnotic experimentation. Since it is necessary to restrict the exposition of facts, we think it well, by way of compensation, to throw light on the questions of method, which constitute the philosophic side of the subject.

The study of hypnotism bristles with difficulties, although this has not occurred to the numerous persons who have expected to find in these questions the occasion of a brilliant and easy success. Although nothing is more simple than the invention of dramatic experiments, which strike the vulgar with fear and astonishment, it is on the other hand very difficult, in many cases, to find the true formula of the experiment which will give its result with convincing accuracy.

Speaking generally, the method is the same in the study of physical phenomena, and in that of the phenomena of suggestion. In order to obtain constant results, which may be verified at pleasure by any other observer, it is necessary to define with the utmost precision the physiological and pathological conditions of the subjects of experiment, and the nature of the processes by which it is performed. Whenever one of these two rules is violated, the method is incorrect. There is a risk of seeing the result announced falsified by another observer, who was unable to reproduce it; hence the questions become confused, personal discussions, which are necessarily sterile, ensue, and finally general disbelief is aroused.

The physical state of the subject must first be defined. We recommend the course pursued by ourselves, of choosing hysterical subjects who display the strongly marked characteristics of profound hypnotism. Any one who wishes to verify the new experiments in suggestion which we adduce, must occupy himself solely with these subjects. It is very important to indicate the physical state of the subject of experiment, and in no other way can there be any comparison of results. It is true that no morbid state is constantly presented under the same aspect. Each individual impresses a peculiar stamp on the morbid state to which he is subject. All diseases are displayed in forms which vary with the constitution of the subject, and it may even be said that each organic function presents individual variations. We cannot, therefore, be surprised that hypnosis is displayed under various and more or less characteristic aspects, but this is an additional reason for only comparing similar facts, unless we wish to fall into deplorable confusion.

It is less easy to define with precision the modes of operation, since the experimenter is often mistaken with respect to the means which he employs. He believes that he is suggesting a given idea, but at the same moment he unconsciously suggests a second, which alters the first idea, or else the subject intervenes in an active manner, in order to simulate certain phenomena, thus deceiving the observer. Simulation and unconscious suggestion are the two rocks to be avoided in studying the facts of suggestion.

Simulation.—It must be admitted that simulation, always a difficulty in the study of hysteria, is no where so formidable as it is in this department of study. The experimenter is safe as long as he has to do with physical phenomena, but this is not the case with many of the facts of suggestion. It is very easy for the subject to simulate an hallucination or delirium. These are internal phenomena which cannot be seen, touched, and handled, like an objective fact; they are subjective phenomena, personal to those who experience them, and consequently they may readily be assumed. Before studying them, it must be proved that they exist. Before observing the characteristics of an artificial hallucination, we must ascertain that it is really experienced by the subject.

The danger is not averted by proving that the subject is really hypnotized, for, as we have already said, simulation and somnambulism do not exclude each other. Pitres has ascertained that even when the subject is asleep she may still deceive. We must, therefore, exact from the facts of suggestion themselves the proof of their reality.

Strictly speaking, we might appeal to moral proofs; but these proofs are only valuable to those who know the subjects; they are strictly personal. Moreover, those who are satisfied with moral proofs should remember

Hublier, whose good faith is undoubted, and who was deceived by his somnambulist subject for four consecutive years. It would be wise to accept the lesson of caution presented by this fact.

The method to be pursued in such cases is already told, and it may be summed up in one word : it is the *experimental method*, and includes the most improved processes of clinical observation and physiological research.

In earlier times magnetizers were content to observe, and many in our day imitate them in this respect. After producing a given psychical phenomenon by suggestion, they observe, and then describe it. We hold that this is only a preparation for the experiment, which remains incomplete. Passive observation is not enough to assure us of the subject's sincerity, nor can the reality of the hallucination suggested be proved, if we only observe what the subject does, and listen to what he says. The investigation must be carried further, and the phenomenon of suggestion must be subjected to oystematic examination, in order to separate it from the objective signs. It is in this way that experiments have shown that hallucinative vision is modified by optical instruments just as actual vision is; that hallucination with respect to a colour produces the same effects of contrast in colour, as if it were actually seen; that artificial anæsthesia produces the same phenomena of colour as the spontaneous achromatopsia of hysterical patients; that the motor paralysis induced by suggestion is accompanied by the same physical signs as a paralysis due to organic causes. These hidden characters, which are revealed by experiment, are evidently of a complex nature; in order to understand them we must be acquainted with physics, psychology, and the physiology of the nervous system. The effects of contrast produced by colour-hallucination is inexplicable, unless we understand the theory of complementary colours. Again, we cannot understand the clinical signs of motor paralysis produced by suggestion, unless we are acquainted with the nature of organic paralysis; and so it is in other cases. There is no risk of the subject's inventing the characteristics as a whole, in order to deceive the experimenter, and we may therefore be assured that there is no simulation, and this for two reasons: first, want of knowledge, and, secondly, want of power. The objective signs mentioned just now are therefore very valuable; they apply to every case, and offer an irrefutable proof of the reality of the experiment.

In short, the method to be pursued with respect to facts of suggestion is this: the artificial psychical phenomena must be matter for experiment, with the aim of rendering these subjective disturbances objective.

A singular problem, however, occurs with reference to simulation, which has not yet been examined by any observer. The rules which we have just laid down in order to counteract simulation are very efficacious when the simulator does not experience in any degree the assumed phenomenon. For instance, if the subject asserts that he has a visual hallucination, when he sees absolutely nothing, the manifold proofs furnished by optical instruments, complementary colours, etc., will easily expose the fraud. But it is open to question whether simulation in a subject liable to suggestion may not effect all which is effected by suggestion itself.

Let us take an important instance. Motor paralysis can be given to some subjects by means of suggestion, and it is a question whether the subject, with the purpose of deceiving the experimenter, may not simulate a motor paralysis, and whether this assumed paralysis may not present the same objective characters as that which is produced by suggestion. We think that this is possible, for in paralysis by suggestion the real cause of functional impotence is the *idea* of a paralysis, and it matters little whether this idea proceeds from suggestion by the experimenter, or from simulation by the subject; it is only essential that it should be sufficiently intense as to affect the motor power. It is in this way that we hold that simulated phenomena may in some cases be absolutely confounded with real phenomena.

This question of simulation in an individual liable to suggestion is, in fact, only one aspect of a much larger question: that of the action of the will on phenomena of suggestion. It may be asked whether an individual liable to suggestion can voluntarily create, modify, and destroy effects on himself comparable to those developed by suggestion. We are acquainted with facts which enable us to reply in the affirmative. We have seen subjects who could at pleasure, and in the waking state, call up the hallucinative image; when looking attentively at a sheet of white paper, they could cause it to appear red, blue, green, etc., and the colour thus evoked would be sufficiently distinct to give birth in due succession to the complementary colour which the subject could indicate correctly. This remarkable visual phenomenon differs from artificial hallucination in one respect; it requires a voluntary effort, continued for a period varying between twenty seconds and a minute, while a suggested hallucination occurs almost spontaneously. We met with another instance of voluntary suggestion in psychical paralysis. A subject to whom complete paralysis of the arm had been given was able to free herself from it after endeavouring for five minutes to move the paralysed limb.

Unconscious Suggestion.—We have not yet had occasion to speak of this last kind of suggestion, which has a tendency to introduce itself like a parasite into the experimenter's mental suggestions, so as altogether to vitiate the results. It should be known that some hysterical subjects become when hypnotized so sensitive and such delicate re-agents, that no word or gesture escapes their notice; they see, hear, and retain everything, like registering instruments. Suppose that the experimenter has produced a visual hallucination; he then wishes to ascertain whether this sensorial disturbance has produced any modification in the sensibility of the integuments of the eye, which were insensitive before the experiment. Before making the examination, he says to one of his assistants, "I am going to see whether the cornea and conjunctiva have become sensitive." The subject hears what is said, and it may easily, although not certainly, happen that the words act as a direct suggestion of the symptom in question, so that the experimenter runs the risk of taking for an effect of hallucination what is the effect of suggestion. The subject acts in good faith, as well as the experimenter; there is no simulation, and yet there may be a considerable error.

The risk of unconscious suggestion is not found in

the subjects of profound hypnotism in all degrees of hypnosis; it is slighter during lethargy and catalepsy, and does not exist at all in the case of some subjects, who are in these states unable to receive any suggestions whatever. It is during somnambulism that unconscious suggestion is most frequently present, and when the experimenter has to do with a somnambulist, he should always remember this source of error, and guard against it. The moral proof derived from the subject's good faith is useless, since we are not concerned with simulation. It is well for the experimenter to work in silence, not to prepare his experiments in the presence of his subject, and to execute them before a limited number of spectators. One or two are enough. We cannot too often repeat that only the first experiments are convincing, since, strictly speaking, these alone are performed on a virgin subject, safe from unconscious suggestion. Every time that the experiment is repeated, there are probably more spectators who comment upon it aloud; they unconsciously make suggestions which vitiate the purity of the phenomenon, and greatly diminish its value. In addition to this there is another source of error, namely, that when the second experiment takes place, the subject remembers the former one. For instance, if a given phenomenon has been produced once by the employment of a given agent, on the second occasion the presence of this agent, or even its image, may recall its sensation, and so disturb the experiment in hand.

For these reasons, among others, we have always been careful in our papers on hypnotism to give the results of the first experiment, although these results were often less exact and complete than those which followed.

We believe that the experimenter who follows the rules just laid down, who accurately defines the physical and mental condition of his subjects, who takes measures completely to eliminate simulation and unconscious suggestion, will obtain verifiable results.

We must not omit a final precept, equally applicable to the research into facts and to the performance of experiments, and that is, to bring together the phenomena of suggestion which are already known, and which make part of positive science. Many experimenters have disregarded this important precept, and have written pages on suggestion which are only a collection of amusing anecdotes, adapted rather to pique curiosity than to afford instruction. Paul Janet has forcibly remarked on the serious consequences of this omission : " In recent works on the subject of suggestion, all more or less intended for the public, since they were published at conferences, lectures, or in reviews, we have observed that, instead of first relying on the most common and elementary facts, assuming that these were already too well known, which is by no means the case, they have been chiefly anxious to dwell on extraordinary facts which strike the imagination. This is intelligible enough, since in addressing the public, success is the first object. The writer doubtless loves truth for its own sake, but he is not unwilling to make it effective. If the audience or the reader is prepared, the effect is weakened, and it becomes greater in proportion to its unexpectedness. This tendency to throw into relief the extraordinary and the unexpected is excellent from the literary and dramatic point of view, but it has many inconveniences when we are concerned with science, for when the amazement is too great, it inclines the mind to unbelief, and diverts it from examination. Enlightened minds have long held aloof from the study of magnetism, precisely because of its marvellous and mysterious character. And now, although the new facts rest, or appear to rest, on a really scientific method, yet their resemblance to those of magnetism tends to produce an analogous disposition to hostility and dislike. At the same time, by a reciprocal and contrary effect which is no less vexatious, others regard these singular phenomena, of which they cannot divine the cause, as if they were invested with the same prestige of the unknown and the mysterious as those of magnetism. The one leads to the other, and since the public is unacquainted with the methods of science, they confuse the subjects together, so as to fall back into the error which it was sought to avoid." *

The method to be pursued consists, as we have said, in at once showing that suggestion is not a distinct phenomenon in the history of intelligence, an isolated, disconnected fact, explained by nothing, and as it were suspended in mid-air. It is necessary to insist on the close relations existing between the phenomena of suggestion and the admitted facts which form part of positive science; the connection between them must be made clear, since these phenomena are only an exaggeration and a pathological deviation.

On this subject we shall have to point out numerous parallels between the facts of suggestion and those of physiology, of psychology, and of mental disorder. The comparative study of suggestion and of psycho-

* *Revue politique et littéraire*, August, 1884.

logical phenomena will show us that the hypnotic subject is not governed by special psychological laws, and that the germs of all his symptoms may be traced in the normal state. The comparative study of the phenomena of suggestion and of mental disorder will, moreover, show that the psychical disturbance caused in the subject by suggestion has many characters in common with the spontaneous disturbance found in an insane person; that, for instance, the hallucination of hypnotism does not essentially differ from the ordinary forms of hallucination.

It is by means of these repeated comparisons that the experimenter finds his bearings in the study of hypnotism, in which such care is necessary. Reliance on the achieved results of positive science acts as a check and guidance, and hypnotism, instead of being merely an amusement for the idle, becomes a useful method of experiment in psychology and in mental diseases.

III.

Suggestion acts on the subject's nervous system, and produces modifications analogous to those which are produced by peripheral excitements. But we are far from knowing all the effects produced by the idea which is introduced by means of suggestion into the subject's brain; it is, indeed, probable that we are not acquainted with the thousandth part of them. Far from wishing to conceal this incompleteness of the theory of suggestion, we think it well to call particular attention to it. The study of suggestion has only just begun, and many surprises are doubtless in store for us.

We think it probable that when suggestion is employed in the case of a subject adapted for it, it is capable of producing all the actions connected with the nervous system. This is, however, only a probability, since direct proofs are wanting, and if it were proved, we have still to learn the extent and limits of the influence of the nervous system on the rest of the organism. The question of suggestion merges in this case in an important and still somewhat obscure question of physiology.

A fresh chapter in the history of suggestion has been lately opened. Various observers have been studying suggestions which do not exert any action on the subject's physical life, but on the so-called vegetative functions, circulation, calorification, secretion, digestion, etc.

We do not propose to dwell on well-known facts, such as purging by means of suggestion, since such facts present no special feature, and we know that in the normal state these effects are produced by certain forms of mental emotion.

The most important of the organic disturbances produced by an idea is an experiment on vesication, performed by Focachon, a chemist at Charmes. He applied some postage stamps to the left shoulder of a hypnotized subject, keeping them in their place with some strips of diachylon and a compress; at the same time he suggested to the subject that he had applied a blister. The subject was watched, and when twenty hours had elapsed the dressing, which had remained untouched, was removed. The epidermis to which it had been applied was thickened and dead and of a yellowish white colour, and this region of the skin was puffy, and surrounded by an intensely red zone.　Several physicians, including Beaunis, confirmed this observation, and the latter made photographs of the blister, which he presented to the Society of Physiological Psychology on June 29, 1885.*

Shortly afterwards, on July 11, 1885, Dumontpallier informed the *Société de Biologie* of experiments performed on hypnotized hysterical subjects, whose temperature he had raised locally several degrees; this is a curious fact, to which another is analogous, namely, that the temperature is lowered in the correlative region of the body during the suggestion of these physical phenomena.

During the same *séance*, Bourru and Burot, professors of the Rochefort school, published records of epistaxis, and even of blood-sweat, produced by suggestion in a male hysterical patient, who was affected by hemiplegia and hemi-anæsthesia.　On one occasion, after one of these experimenters had hypnotized the subject, he traced his name with the blunted end of a probe on both his fore-arms, and then issued the following order:—"This afternoon, at four o'clock, you will go to sleep, and blood will then issue from your arms on the lines which I have now traced."　The subject fell asleep at the hour named; the letters then appeared on his left arm, marked in relief, and of a bright red colour which contrasted with the general paleness of the skin, and there were even minute drops of blood in several places.　There was

* It seems that as long ago as November, 1840, Prejalmini, an Italian physician, raised a blister by applying to the healthy skin of a somnambulist a piece of paper on which he had written a prescription for a blister. We owe this fact to Ferrari, who found it in Ricard's *Journal de Magnetisme Animal*, 2nd year, 1840, pp. 18, 151.

absolutely nothing to be seen on the right and paralysed side. Mabille subsequently heard the same patient, in a spontaneous attack of hysteria, command his arm to bleed, and soon afterwards the cutaneous hæmorrhage just described was displayed. These strange phenomena recall, and also explain, the bleeding stigmata which have been repeatedly observed in the subjects of religious ecstasy who have pictured to themselves the passion of Christ.

Charcot and his pupils at the Salpêtrière have often produced the effects of burns upon the skin of hypnotized subjects by means of suggestion. The idea of the burn does not take effect immediately, but after the lapse of some hours. It is still very doubtful whether all organic functions may be thus modified by means of suggestion.

Quite recently one of the present writers * succeeded in showing, by means of processes analogous to those of Mosso, that any part of the body of an hysterical patient may change in volume, simply owing to the fact that the patient's attention is fixed on that part. This important observation is not only an addition to the foregoing, but explains them by showing what influence may be exerted in hyperexcitable subjects by a simple phenomenon of ideation on the vaso-motor centres, which are concerned in all experiments of this kind.

Among the effects of suggestion only one class has been the object of regular research: that of psychical phenomena. These have been studied by preference because they were the first which charlatans sought to turn to their advantage. To these we now propose to turn

* Ch. Féré, *Bull. Société de Biologie*, 1886, p. 309.

our attention, endeavouring to define and classify them with the utmost care.

If, before going into details, we consider the subject as a whole, it will be seen that we have to study the part played by ideas in the modification of the intelligence; that we must observe what this factor produces when it acts alone. It is generally admitted by psychologists that the idea is only a secondary factor, that it is for the most part a resultant, a point of arrival; that psychical phenomena are in some sort developed from below in an upward direction. They do not begin in the upper centres of ideation, but are completed there.* So also the phenomena of suggestion which, by an inverse mechanism, are developed from the higher to the lower plane, are more superficial and ephemeral than spontaneous phenomena. It is possible to suggest to a subject that he is very hungry, but this feeling, which is dictated by an idea, will not be so deep as that which is due to an organic necessity. So, again, the personality of the subject may be transformed, and he may be changed into a dog or a wolf; but when this borrowed personality is engrafted on the true one, the character is not fundamentally changed. If the suggestion is to produce any permanent modifications it must be often repeated, and it will then, at least in some cases, end by producing habits. A subject to whom suggestions of motor paralysis had been repeatedly made, said that in dreams she often saw half of her body paralysed. Experimental suggestions of crime ought not to be lightly made, since we cannot always tell what traces they leave behind them.

* Ribot, *Maladies de la Personnalité*, p. 131. Paris: F. Alcan.

A careful study of suggestion shows that this word does not imply a single fact, but two principal facts which may be said to form the cardinal points of the whole theory. There are two fundamental kinds of suggestion; the one produces an active or impulsive phenomenon, such as a sensation of pain, an hallucination, an act; the effect of the other is to produce a phenomenon of paralysis, such as the flaccidity of a limb, the loss of memory, anæsthesia of the senses. These are two quite different processes, and may even be said to be opposed to each other, since the one undoes what is done by the other. It is impossible to refer them to the same psychical law, and to apply to them the same explanation.

Let us first consider the positive suggestions, of which the hallucination and the act are the most important. It has already been observed that all suggestions are addressed to the subject's sensorial organs, and that the co-operation of his intelligence is necessary in order to attain the end in view. We must go further, and establish the fact that each suggestion includes two things : first, an impression is made upon the subject, which is, according to circumstances, a sensation of sight, of hearing (verbal suggestion), of touch, or of the other senses. This initial impression, which may be termed the suggestive impression, has the effect of arousing in the subject's brain a second impression, which may be termed the suggested state, such, for instance, as the hallucinatory image. The first impression is the means, the second its object. In reply to the question how the first impression, which is directly produced by the experimenter, can arouse the second, which is wholly from within, and without any direct influence from the experimenter, we should reply that this is owing to the association of ideas. Suggestion in its positive form is only the setting in action of a mental association previously existing in the subject's mind.

Suppose, for example, that the subject is told to look at the bird on her apron. As soon as the words are uttered she sees the bird, she feels it in her hand, and can sometimes even hear it sing. Inexperienced persons may think it extraordinary and even inexplicable that an imaginary image can be created in the subject's brain by mere words; but it is nevertheless a fact that the association of ideas is the cause of the suggestion of hallucinations. The words uttered by the experimenter are associated with the mental image of a bird by education, by repetition, in a word, by habit; therefore it produces this image, and the hallucination is effected. It is a law that when two images have frequently been received together, simultaneously or in quick succession, the presence of the one tends to revive the other. The production of the hallucinatory image by means of verbal affirmation is only a fulfilment of this well-known law. We should be able to show, by considering it more closely, that this mode of suggestion belongs to the group of associations by proximity.

Instead of making use of speech, the subject's sense of sight may be employed. When his eyes are mobile and follow all our actions with docility, we make with the hand the appearance of some flying object, and the subject soon exclaims, " What a beautiful bird ! " This singular effect of a simple gesture is also due to the association of ideas. The rude imitation by which the hand represents the movement of some flying object has raised up the image of a bird. In this case the association which comes into play differs from the former one; it is an association due to resemblance.

It is in this way that psychology explains the mechanism of hallucinatory suggestion, which essentially consists in acting on an association. It is only a special case of the great law of which this is the formula :—When an image is aroused in the mind, it tends to reproduce all the images which resemble it, or which were found with it in an anterior act of consciousness. In a word, one image suggests another. Paul Janet observes on this subject: " Some Scotch psychologists, Brown for example, have even proposed to call this law the law of suggestion, a term which would be much more appropriate than the other. I have no doubt that the expression of suggestion, introduced by Braid into his theory of hypnotism, is derived from this source." *

Just as the association of the word with the image explains the suggestion of hallucinations, so the association of the image with the movement explains the suggestion of acts.

When a given movement is executed before the eyes of a subject, such as clapping the hands, the representation of this movement is produced in his mind. When, without moving himself, the experimenter bids his subject clap his hands, the representation of the same action is aroused in his mind by the association of words with ideas ; if in both cases alike the subject performs the act in question—in other words, if the image is translated into movement—this is because custom has associated the image with the movement. It is, as it

* *Revue politique et littéraire,* August, 1884.

were, the beginning or first stage of the movement which it represents, and owing to this fact the subject automatically executes whatever he is ordered to do, even if the act should be dangerous, immoral, or merely ridiculous. Richet tells us that on one occasion, when performing experiments on the friend he had hypnotized, he compelled him to pick up the piece of chalk he had thrown under the table twenty times. In fact, the suggestion of acts is perhaps, of all the phenomena of suggestion, the one which approximates most closely to the normal state; it simply consists in the servile execution of an order.

We have, however, an observation to make on this subject. We are compelled to connect these facts of suggestion, impulses, and hallucinations, with the facts of positive science which may serve to prove and control them. Yet we are far from believing that this method will give a complete explanation of the phenomena in question. It is a proof of excessive confidence to assume that everything is explained, and a word is enough to show that the matter is still full of obscurity. Admitting that the suggestion of a movement is explained by the association of the movement with its image or representation, it is a question whether as much can be said of the suggestion of an act. When the subject's brain is charged with this idea, "On awaking you will steal the handkerchief of some given person," and the subject when he awakes does actually commit the theft, we cannot suppose that there is nothing in this except an image associated with an act. The subject has, in fact, appropriated and assimilated the experimenter's idea. Instead of passively executing the order given by another, the order has passed into the active state, that is, the subject feels a desire to steal—a complex and obscure state which no one has hitherto been able to explain. We shall return to this subject presently, in order to examine it more closely.

Since we find so much that is enigmatical in the region of impulsive suggestions, which are the clearest and most intelligible, this is still more the case when we approach the subject of inhibitory suggestions. Here the most superficial psychologist will find himself on new ground. The facts of paralysis by means of suggestion completely overthrow classic psychology. The experimenter who produces them with perfect ease does not really know what he is doing.

Take an instance of systematic anæsthesia. The subject was told, "On awaking you will be unable to see or hear, or in any way perceive M. X——, who is now present; he will have completely disappeared." Accordingly, when the subject awoke she saw all the persons who surrounded her with the exception of M. X——. When he spoke, she did not answer his questions, and when he laid his hand on her shoulder, she was unconscious of the contact. He put himself in her way, and she walked on and was alarmed to encounter an invisible object. We are ignorant how this phenomenon is produced and can only accept the external fact; namely, that when a subject is assured that an object present has no existence, the suggestion has the direct or indirect effect of establishing in his brain an anæsthesia corresponding to the object selected. But it is still a question what occurs between the spoken affirmation, which is the means, and the systematic anæsthesia, which is the end. We cannot, as in the case of hallucination, assume that the word spoken to the subject, and the phenomenon produced, are connected by association. If it is true that the image of a serpent is associated with the words, " Here is a serpent," it cannot be said that the incapacity for seeing M. X——, who is present at the time, is also associated with the words, " M. X—— is non-existent." In this case the law of association, which is so useful in resolving psychological problems, is altogether unavailing. This is probably because this law will not explain all the facts of consciousness, and is less general than it is supposed to be by English psychologists.

Similar reflections apply to another instance of paralysis by suggestion, or motor paralysis. It is possible to suggest to a hypnotized subject that her arm is paralysed. It is only necessary to say repeatedly with the requisite authority, " Your arm is paralysed," and functional impotence is soon displayed. The subject begins by signifying a denial of the fact, she tries to raise her arm, and succeeds in doing so. She is repeatedly told, " You cannot raise your arm; it will fall again," and paralysis gradually comes on, and presently extends to the whole arm. The subject can no longer move it, and its flaccidity is absolute. Such is the singular phenomenon of motor paralysis by means of suggestion. It is as difficult to understand as the anæsthesia to which it corresponds. We do not think that it can be explained by any psychical facts now known to us.

Perhaps this whole class of facts is subject to a general psychical law of which the most advanced psychologists have not yet discovered the formula and which may have some analogy with an inhibitory action.

According to this hypothesis it may be provisionally admitted that in order to paralyse the subject the experimenter produces in him a mental impression which has an inhibitory effect on one of his sensorial or motor functions: it should also be clearly understood that, strictly speaking, it is not the mental impression which produces the inhibition, but the concomitant physiological process. Moreover, it must be remembered that the word inhibition explains nothing, and does not dispense with the necessity of seeking the true explanation.

Hypnotic suggestions may be classified as follows:— Some are only effected during sleep, and disappear with a return to the normal state; others continue during the waking state; others, again, are produced in the waking state.

Thus, the hallucination of a bird may be given to a somnambulist, and this hallucination will disappear when the subject is awakened by breathing on his eyes. As soon as he returns to the normal state, he is completely free from any imaginary vision. This is the case with fresh subjects and with those who are only moderately receptive, in whom suggestion does not outlast the hypnotic sleep. It is possible, however, to protract the suggestion after the awakening by strengthening it with a different suggestion. If the precaution is taken of telling the subject to whom the hallucination is given that he will still see the object in question when he awakes, the assertion is often enough to ensure the existence of the suggestion in the post-hypnotic state.

It is not usually necessary to make a special suggestion in order to produce this effect in subjects thoroughly under the influence of profound hypnotism. Every effect which has been suggested without fixing any term to it, and which has a continuous form, may persist for a shorter or longer period during the waking state. This is especially the case with hallucinations, paralysis, etc. We are thus presented with a curious experience, calculated to interest the psychologist. The subject is awake, has returned to what may be called his normal state, is able to reflect, reason, and direct his conduct; and yet, under these conditions, he is influenced by the hypnotic suggestion.

The suggestion which persists during the waking state presents one interesting characteristic; it appears to the subject to be spontaneous. As a general rule, the process which produced the suggestion seems to leave no trace of its symptoms, and the subject who after awaking is still a prey to the hallucination which was suggested, does not remember the way in which it was produced. We have not in a single case met with a subject who said spontaneously, "If I see a bird at this moment, it is because you assured me that I saw one when I was asleep." The memory of the uttered word has completely disappeared, while its effect remains in the hallucinatory image. Hence it follows that hypnotic hallucination has always the appearance of a spontaneous symptom. Some curious consequences ensue. A subject is told that one of the persons present wears a coat with gold buttons, and the word arouses the sensible image of buttons of a yellow colour. If the subject is afterwards asked of what the buttons are made, he may reply, "They are made of copper." The buttons which he has before his eyes are yellow, and he supposes that they are copper; he has completely forgotten the word *gold* which figured in the verbal suggestion.

The same may be said with respect to suggestion of acts. On awaking, the subject obediently performs the act which he was ordered to do during the hypnotic sleep, but he does not remember who gave him the order, nor even that it was given at all. If asked why he is performing this act, he usually replies that he does not know, or that the idea has come into his head. He generally supposes it to be a spontaneous act, and sometimes he even invents reasons to explain his conduct. All this shows that the memory of the suggestion, so far as respects its utterance, is completely effaced.

The same rule applies to paralysis by suggestion. The subject who is affected by a monoplegia, when he awakes does not understand how the accident occurred; he remembers nothing of the verbal suggestion, nor does he suspect that the incapacity to move his arm is due to a conviction of the want of motor power. In short, the suggestion is effaced from the subject's mind as soon as it has produced its effect, and the symptoms appear to be evolved independently of their cause. The existence of this partial failure of memory may perhaps allow us to compare the artificial results of suggestion with the phenomena which are spontaneously displayed in normal individuals and also in the insane, in their acts, phenomenal impulses, hallucinations, etc.

We now come to the detailed study of the facts of suggestion. It is impossible to examine all of them, and we must be content with selecting a certain number of typical cases for careful study. We shall successively consider hallucinations, impulsive acts, motor paralysis, and paralysis of the senses. The phenomena just indicated are the simplest which can be obtained by means of hypnotic experimentation, and in studying them we may be said to study the most elementary properties of the phenomena of suggestion. If space permitted, we should follow up the study of these elementary facts of suggestion by enumerating the complex facts derived from them. Thus, we might connect with the hallucinatory image all the facts included under the name of *intelligence*—sensation, the association of images, memory, reason, and imagination. With a suggested act are connected sentiments, emotions, passions, voluntary action, and all the phenomena constituting the psychology of movement, with which we are as yet imperfectly acquainted. Finally, paralysis by suggestion is connected with the phenomena of psychical inhibition of which the study has not as yet even begun.

CHAPTER IX.

I.

HYPNOTIC hallucination, of which we propose to give a short sketch, is undoubtedly one of the most important phenomena of hypnosis; the attention of observers has long been directed to it, and it has been the subject of numerous experiments.

In the case of a subject sensitive to suggestion, the experimenter can produce the most varied hallucinations, and it may almost be said that there is nothing which suggestion cannot create. This observation will suffice, and we need not cite the innumerable instances of hallucinations given by some authors, who are more interested in experiments which amuse than in those which instruct. It is as unprofitable to enumerate all the species of hallucination which it pleases the observer to impose upon his subject, as to describe all the forms which a piece of clay may assume in the hands of its moulder. We shall, therefore, content ourselves with giving a few instances of the way in which hypnotic hallucination may affect all the senses.

Sight.—A false appreciation of the form of an object may be suggested, so that it appears to the subject to be larger, smaller, or misshapen. If, for example, the idea of some deformity of face in a given person is suggested, the subject, even when some hours have elapsed since his awakening, will regard that person with an expression of disgust or horror whenever he looks that way, and, indeed, the person in question may sometimes become an object of permanent dislike. We have employed this method with success in order to break off the relations between certain hysterical patients. The illusion may be carried so far as to produce a mistake with respect to the identity of a person; an hypnotic subject will in the waking state lavish caresses on a person whom she is known to detest, if during the hypnotic sleep it has been suggested to her that she has to do with some other person to whom she is attached, and the error will sometimes persist for a whole day, until the illusion is destroyed by natural sleep or by an hysterical attack. If the presence of a person who is really absent has been evoked during the hypnotic sleep, the illusion is equally persistent, and the subject may perceive an imaginary object throughout the day. At the word of the experimenter the laboratory becomes a street, a garden, a cemetery, a lake, etc; a portrait appears on a blank sheet of paper. It may even be suggested that there is a column of figures on the paper, which the subject will add up correctly. (Babinski.)

Hearing.—Influenced by suggestion, the subject confounds the voice of an unknown person with that of an absent acquaintance; he can hear, in the midst of profound silence, voices which issue orders, which repeat insults or obscene words, etc.

Taste.—If the subject is presented with a piece of

paper, and told that it is a cake, he will begin to eat it with relish. In other cases, he may be convinced that his food is poisoned. If the idea of a nauseous substance is suggested to him, the sensation may be so intense as to produce vomiting.

Smell.—This sense may also become the seat of erroneous impressions. The subject may, for example, believe that a bad smell is coming to him through the keyhole, etc.

Touch.—The illusions and hallucinations of touch assume still more varied forms, and all forms of cutaneous sensibility may occur together or separately. The suggestion of a wound is one of the most curious of these hallucinations; the subject's description of his suffering varies with the suggestion that the wound was given by a sharp or blunt instrument, but his description is only correct if he has previously experienced one of these accidents. It is still more remarkable that the hallucination of sight is simultaneously developed : the subject sees the blood flow, etc., and a systematic delirium ensues which is more or less persistent, and during which he complains of imaginary suffering, applies appropriate dressings, and carries his arm in a sling, just as if the wound really existed.

Muscular Sense.—If an hallucinatory object, such as a lamp-shade, is put in the subject's hands, and he is told to press it, he experiences a sensation of resistance, and is unable to bring his hands together.

Internal Sense.—Suggestion cannot only be applied to the senses; it is possible to produce visceral hallucinations and illusions, the sensation of a foreign substance in the interior of the body, etc. But the most remarkable of this group of suggestions, and the one which it is the most easy to produce, are those which refer to the calls of nature. When hunger or thirst has been suggested, the subject eagerly calls for food or drink as soon as he awakes, and if they are presented to him, he swallows them with avidity. If it is suggested that he wishes to make water, the subject is scarcely awake before he assumes an embarrassing attitude and hastens to satisfy the imaginary necessity. Sensual suggestions provoke equally imperious desires, of which the consequences may be imagined.

Nor is this all. Not only do the suggestions of imaginary sensations affect the senses and the viscera, but it is possible to suggest the idea that there is a change of structure in the whole substance. The subject may, for instance, awake in amazement, saying, "I am made of glass, do not touch me," and systematic delirium may ensue from this mistaken idea. Other forms of delirium may be created at pleasure, by the suggestion of a sensation affecting one of the special senses. Richet's observations on this subject, in the *Revue Philosophique* of March, 1884, are worthy of notice, and we subjoin a few of them.

Mme. A——, a respectable matron, underwent the following metamorphoses:—*As a peasant.* She rubbed her eyes and stretched herself: "What o'clock is it? Four in the morning!" She drags her feet as if wearing sabots. "I must get up and go to the stable. Now, La Rousse, turn round!" She assumes to be milking a cow. "Leave me alone, Gros-Jean; leave me alone, I say, and let me get on with my work." *As an actress.* Her face, so harsh and dissatisfied a moment before, assumes a smiling expression. " You see my skirt ? My director insisted that it should be longer. In my opinion, the shorter the better; but these directors are always annoying. Do come and see me sometimes; I am always at home at three. You might pay me a visit, and bring a present with you." *As Archbishop of Paris.* Her face assumes a very serious expression, and she speaks slowly, in a voice sweet as honey: "I must finish writing my charge. Oh, it is you, M. le grand vicaire. What do you want ? I did not wish to be disturbed. . . . Yes, this is New Year's Day, and I must go to the cathedral. . . . This is a very reverent crowd, is it not, M. le grand vicaire ? There is still a sense of religion in the people, whatever happens. Let that child come near, that I may bless him." She presents an imaginary ring for the child to kiss, and throughout this scene she makes gestures of benediction to the right and left. "I have now another task in hand. I must go and pay my respects to the President of the Republic. M. le President, I give you my good wishes. The Church wishes you a long life: in spite of the cruel attacks made upon her, she knows that she has nothing to fear as long as a perfectly honest man is at the head of the Republic." She pauses, appears to listen, and says aside, "Yes, yes, only false promises !" Then aloud, "Now let us pray;" and she kneels down.

We have observed some phenomena of the same kind, but in a less developed form. On one occasion we told X—— that she had become M. F——, and after some resistance she accepted the suggestion. On awaking she was unable to see M. F——, who was present; she imitated his manner, and made the gesture of putting both her hands in the pockets of an imaginary hospital apron. From time to time she put her hand to her lips, as if to smooth her moustache, and looked about her with assurance. But she said nothing. We asked her whether she was acquainted with X——. She hesitated for a moment, and then replied, with a contemptuous shrug of the shoulders, " Oh yes, an hysterical patient. What do you think of her ? She is not too wise."

It is difficult to define the psychical nature of these transformations of personality. In our opinion the phenomenon is more complex than that of hallucination, and constitutes true delirium. Moreover, many hypnotic hallucinations—that of hearing, for instance—have a secondary tendency to produce a delirium corresponding with their character.

The form of hallucinatory suggestion may be varied. We have begun by considering the hallucinations which relate to the present time, those which are realized as soon as the suggestion is given. It is possible, in the case of some subjects, to create an hallucination which is to be realized in a given number of days, weeks, or even months. A simple affirmation is enough to effect this experiment. The patient is told that when he enters the room on the following day he will see a crow perched on the table ; or that two months hence, on the 1st of January, he will see the speaker come in to wish him a happy new year. The subject remembers nothing of this when he awakes, and the suggestion remains dormant in his mind until the date fixed for its revival. We shall have more to say on the subject of these experiments.

On the other hand, retrospective hallucinations, which are really hallucinations of the memory, can also be given. It is, for instance, impressed upon the subject that at a given moment of his past life he witnessed the commission of a crime by an old man living in the same house with him (Bernheim), and if the suggestion is clearly defined, the subject's memory will be as intense and as full of details as if the fact had actually occurred. We can see what grave consequences might ensue from these experiments from a medico-legal point of view.

Unilateral Hallucinations. — The hallucinations hitherto in question are bi-lateral; the senses all agree to deceive the subject: what the eye sees, that the hand touches, and the ear hears. By means of suggestion, however, a subject can receive a uni-lateral hallucination, as when an imaginary object is presented to him which he can only see with one eye. Dumontpallier was the first to study this phenomenon, which is common in the insane, and may be produced in several ways. For instance, the subject is told that there is a portrait on a blank sheet of paper, and it is added, after opening the right eye only, " You see this portrait ? " Then this eye is closed and the other is opened, with the words, " You no longer see anything." On awaking the hallucination remains, localized in the right eye, with which the subject sees the portrait, while for the left eye the paper remains blank. The experiment, performed in this way, is simple enough. Dumontpallier has made it more complex by giving different hallucinations to each of the two symmetrical organs, to each eye or each ear. Thus, after hypnotizing the subject, he says to the right ear that it is a fine, sunshiny day, while another person says to the left ear that it is raining. On the right side of the subject's face there is a smile, while on the left the lip is drawn down, so as to display annoyance at the bad weather. The experiment is continued by the intervention of sight and hearing, and the description of a rustic *fête*, attended by young people of both sexes, is transmitted to the right ear. This description is perceived by the left cerebral hemisphere, as it appears from the smile on the right side of the face, while on the left there is an expression of emotion, caused by the imitation of the barking of a dog at that ear. It is said that the different expression of the two halves of the face is most striking. These bilateral hallucinations, which may also be sometimes observed in the insane, are very interesting from a psychological point of view. Dumontpallier thinks that they may be taken as a proof of the functional independence of the two cerebral hemispheres.[*]

In connection with this order of ideas, we will adduce a fresh fact, which we have repeatedly observed. Suppose the idea is given to the subject that a white cardboard appears red to the right eye only; if he closes the left eye while looking at the cardboard with the right eye, it appears to be of a brilliant red; if he uses both eyes, the colour appears to be pink. It is probable that the sensation of whiteness received by the left eye exerts an attenuating effect on the hallucination of the right eye, and thus produces the degradation of colour. The two following facts are connected with this experiment :—If one eye is fixed on a red square, and the other on a white surface, the sensation of red persists, but it is eclipsed from time to time as if by a white cloud. If a red image is produced in one eye after gazing fixedly at

* * *

[*] *Société de Biologie,* 1882, p. 786; Bérillon, *De l'indépendance fonctionelle des deux hémispheres cérébraux* (*Thèse de Paris,* 1884, p. 175).

a green square, and the other eye is then opened on a white surface, the consecutive monocular image is soon effaced. The experiment of the unilateral hallucination of colour holds a middle place between these two; the hallucinatory red image is weakened by the sensation of whiteness received by the other eye, but it is not so much weakened as the consecutive image, and it is more weakened than the actual sensation. With the exception of these differences in intensity, the three phenomena may be referred to a single fact, belonging to the study of optics, and termed the concurrence or struggle of the two fields of vision.

Nor is this all. The result is complicated if it is suggested that the cardboard appears red to the right eye, and green to the left eye. In the subjects observed by us, we have not found that there was a confusion of the two suggested colours, but a species of conflict: at first the cardboard appeared to be red, and a moment after it was green, and this alternation of colour seemed to perplex and weary the patient's sight. This second experiment may be explained like the former one, by the struggle between the two fields of vision. In the normal state we find that this struggle occurs when two different colours, such as red and blue, are simultaneously presented to the right and left eyes. The subject does not, as might have been supposed, see a composite colour, but a kind of mist floating over both colours, and occasionally displacing them. Finally, on looking through the stereoscope at two similar images, one white and the other black, the colour of the images is not fused into a uniform grey, but a conflict takes place between the two fields of vision, so that at one time the bright, at the other the obscure, shade predominates, and hence there results the impression of a shining surface.*   These well-known phenomena seem to explain the experiments on hallucinations which we have mentioned above.

Before going further, we must draw some psychological conclusions from the facts just enumerated. Most modern psychologists accept the law indicated by Dugald Stewart, and more regularly developed by Taine,† according to which every image involves a momentary belief in the reality of its object. Dugald Stewart observed that few men could look from the top of a high tower without experiencing a sensation of fear, although their reason convinces them that they run no greater danger than if they were standing on the ground. "In fact," as Taine adds, "on looking suddenly down, we imagine ourselves to be suddenly thrown headlong to the bottom, and this imagination only terrifies us, because for an imperceptible moment of time it is a belief. We instinctively draw back, as if we felt ourselves falling." In every image presented to the mind there is, therefore, the germ of an hallucination, which only needs development. Such development occurs in the hypnotic state, in which it is only necessary to name a given object to the subject, simply to say, "Here is a bird!" in order that the image suggested by the experimenter's words should become an hallucination. Thus there is only a difference of degree between the idea of an object, and the hallucination of that object.

There is one striking fact to be noted, namely, that

* Bernstein, *On the Senses.* For greater details, see Holmholtz, *Optique physiologique,* p. 964.

† *De l'Intelligence,* vol. i. p. 89.

most of the patients whom we have employed as subjects
of experiments in hallucination are, in the waking state,
endowed with a special power of representing objects in
a sensible form. Liébault regards this quality as the
sign of individuals susceptible of hypnotism. Without
fully accepting this opinion, we believe that persons who
have what Galton terms the power of *visualizing*, are
more susceptible of visual hallucinations than others.
When we request one of our subjects to picture to himself an absent person, he soon declares that he sees that
person as distinctly as if he were actually present. This
vivid power of representation is frequently found in
hysterical patients, and it explains why, when such
patients are gathered together, they, by exchanging
confidences or by imparting their respective impressions,
reciprocally hallucinate each other.

When susceptible and hysterical subjects have been
hypnotized by the same experimenter for several days,
they often end by remaining in a state of permanent
obsession ; they are, so to speak, *possessed*, both by day
in the waking state, and at night during their dreams.
This state of mind is accompanied by spontaneous hallucinations of varying form, but of which the experimenter
is always the object. One patient will have an incubus,
another will be tormented, embraced, etc. If several
subjects meet under the same conditions, and confidences
are exchanged, a species of epidemic of hysterical delirium
ensues, in which the hallucinations will be followed by
impulses, acts of violence, etc., which would account for
the different phases of the drama which terminated in
the death of Urban Grandier. One of the present writers
was present at a scene of this nature, which showed that

such methods of experiment ought to be conducted with the utmost care.*

As we have just seen, hallucination consists in the vivid external projection of an image. But these terms have the defect of treating the image as a thing, as a unity. Yet reflection soon shows that this assumed unity is composed of numerous and heterogenous elements— that it is an association, a group, a fusion, a complexity, a multiplicity.† The image of a ball is the resultant of complex sensations of sight, touch, and muscular sense. It is, therefore, interesting to know if, when an image is associated by contiguity with several others, the external projection of the first image involves that of the others. This occurs in numerous cases of hallucination which may be referred to the action of the memory. Heidenhain gave a series of hallucinations to a hypnotized student, in which he went to the hippodrome, and then to the *Jardin des Plantes*, where he saw the lions come out of their cages. When, some time afterwards, the subject was again hypnotized, the same series of hallucinations occurred spontaneously. So, again, if the subject was reminded of his normal life, or, rather, if an hallucinatory suggestion was made to him, the memory of subsequent events was evoked in its turn, and formed a tableau or hallucinatory scene. In this way a subject may be constrained to live over again a part of his life, and secrets are revealed which would never have been uttered in the waking state, nor even perhaps under hypnotism. We may, for example, cite the story of the singer given by

* Ch. Féré, *Les hypnotiques hystériques considérées comme sujets d'ex-périence mentale*, etc. (*Societé Medico-psychologique*, 1883).

† Ribot, *Maladies de la mémoire*, p. 15. Paris.

Mesnet. If a curved stick was given to him, which he took for a gun, his military recollections revived; he loaded his gun, crouched down, took aim and fired. If a roll of paper was given to him, and a light was flashed across his eyes, recollections of his present profession of singer at a *café-chantant* were aroused; he unrolled the paper and sang loudly. Finally, if a story is told to the subject, and he is then hypnotized, it is not impossible that as soon as he is put upon the track, he may have an hallucination of all the events in succession, as they were related to him.

This tendency of hallucinatory images to suggest themselves shows that the law of the association of ideas by proximity may be exerted without the participation of the subject's intelligence and will. One image provokes another, in virtue of the bond which unites them, and in the same way the second suggests the third. This is one of the clearest manifestations of cerebral automatism.

In pushing the analysis a little further, it may be observed that in these kinds of hallucination it is not merely the image, taken by itself, which is externally projected, but the bond of association which unites several images. It is, in fact, this association which formulates the hallucination; it produces the successive projection of the images in the order in which they are grouped in the mind. This proves that the law laid down by Dugald Stewart with respect to the states of consciousness also applies to the relations of these states. In reply to the question, What is meant by external projection? we answer that it is the belief in the reality of a thing. The external projection of an image is, therefore, the belief in its reality. So that, if it is true that we are inclined to make an external projection of the associated images existing in the mind, this implies that we are inclined to believe that things are in reality associated together, just as their images are associated in the mind. This is no new idea; a considerable time has elapsed since it was formulated by Stuart Mill, and it is confirmed by the facts of hypnosis in the most striking manner.

Speaking generally, we may say that, whenever two images are in association, an implicit affirmation of the existence of a relation between two things ensues; this is an opinion which, therefore, must be referred to an association of externally projected images.

II.

One of the most striking characteristics of hypnotic hallucination is the permanence of its localization. Take the hallucination of a portrait, which is instructive in many respects. If, by means of suggestion, a portrait is caused to appear on a sheet of cardboard of which both sides are alike, the picture will always be seen on the same side of the cardboard, and in whatever direction it may be presented to him, the subject will always place it in the position which it occupied at the moment of suggestion, so that the picture may not be inverted, nor even inclined. If the cardboard is turned round, the portrait is no longer seen, and if it is turned upside down, the portrait is seen with its head downwards. The subject never makes a mistake; if his eyes are covered, or if the experimenter stands behind him while changing the position of the object, his answers are always in conformity with its original localization.

This fact is still more clearly shown by an experiment devised by one of the present writers.* We place a blank card on a blank sheet of paper, and with a blunt pointer, which does not, however, touch the paper, we follow the outline of the card so as to suggest the idea of a black line. We ask the subject, on awaking, to fold the paper in accordance with these imaginary lines; he holds the paper as far from him as it was at the moment of suggestion, and he folds it so as to form a rectangle which precisely covers the card.

Charcot has often repeated a curious experiment which fundamentally resembles the foregoing one. It is suggested to the subject that there is a portrait on a blank card, which is then mixed with a dozen others which appear to be precisely similar. The subject is requested, on awaking, to run his eye over these cards, which he does without understanding the reason, but when he comes to the card on which the imaginary portrait was traced he sees it at once.

All these experiments seem to imply that the hallucinatory image produced in the subject by verbal suggestion does not remain in his brain in a vague and floating state; it is probable, as one of the present writers has shown,† that this image is associated with some external mark—a dot, for instance, or a raised spot,—some distinctive feature of the blank card which was shown to him when the suggestion was made, and this association of the

---

* Ch. Féré, *Les hypnotiques hystériques considérées comme sujets d'expérience, etc.* Paris, 1883.

† Binet, *L'Hallucination (Revue Philosophique,* April, May, 1884).

cerebral image with an external mark would explain the series of facts of which we have given an account.

One detail of these experiments is significant. If, instead of putting the pack of cards into the subject's hands, we show him the imaginary portrait while holding it two yards from his eyes, the card still appears to him. to be white, although a real photograph would appear to be grey. If the card is gradually brought nearer to his eyes, the imaginary portrait becomes visible, but it must be brought much nearer than an ordinary photograph before the subject can say for whom it is meant. This peculiarity can be explained on the assumption that the hallucinatory image is evoked by distinctive marks on the card which are only visible at a short distance. So again, the subject cannot distinguish the portrait through a sheet of tissue paper placed upon the card. With the help of an opera glass, the subject can recognize the object of hallucination when it is too far for him to perceive it with his naked eye, and the same explanation applies to this experiment, although there is an air of paradox about it.

Without going further into the matter, these observations may be summed up in the following formula: The imaginary object presented by hallucination is perceived under the same conditions as if it were real. This formula has served as our guide in a series of experiments on visual hallucination, which we have endeavoured to modify by optical instruments. We proceed to indicate the most important results of these researches into what Janet terms hallucinatory optics.

The origin of these researches is found in an early experiment by Brewster, which was performed in the following way:—It is well known that if, in the normal state, a finger is pressed upon the eye so as to divert it from its normal position, and at the same time the individual looks fixedly at some external object, his sight is doubled, and he sees two objects instead of one. Brewster performed this experiment on a subject who had visual hallucinations, and pressure on the eye caused a duplication of the imaginary object. Paterson tells us that this curious experiment was repeated in analogous circumstances by a student subject to hallucinations. When this student was crossing a garden, he perceived a phantom, wrapped in a large blue cloak, standing a little way off under a tree. The student desired to verify Brewster's famous experiment, and pressed one eyeball, which only had the effect of rendering the figure less distinct. But on looking obliquely he saw the same figure double, and of the natural size. This observation has been confirmed by others. Ball mentions an hysterical female servant who was subject to ecstatic crises in which the Virgin appeared to her in a splendid robe; pressure on the eyeball caused a duplication of the vision, and she beheld two Virgins.

This experiment of duplication has served as the point of departure for researches intended to establish the reality of the subjective phenomena produced in hypnotic subjects.* It occurred to one of the present writers to substitute a prism for pressure on the eyeball. If, when regarding external objects, a prism is placed before one eye, the objects appear to be double, and one of the images presents a deviation of which the direction

* Féré, *Mouvements de la pupille et propriétés du prisme dans les hallucinations provoquées des hystériques* (*Soc. de Biologie*, December, 1881).

and extent may be calculated. If, during the hypnotic sleep, it is suggested to the subject that a profile portrait is on a table of dark wood before him, he distinctly sees this portrait on awaking. If, without warning, a prism is placed before one eye, the subject is astonished to see two portraits, and the position of the false image is always in conformity with the laws of physics. Two of our subjects answered correctly in the cataleptic state, although they were unacquainted with the properties of the prism. Moreover, by concealing its edges they may be deceived as to the precise position in which it is placed. If the base of the prism is uppermost, the two images are placed one below the other, and if the base is lateral, the images are side by side. Finally, the table may be brought so near that the duplication of the image ceases, and this will serve as an index.*

This experiment with the prism is only a variation of the one performed by Brewster. The prism, as well as ocular pressure, exerts two distinct actions on the hallucinatory image—that of duplication and of deviation. Deviation by means of the prism is a more accurate phenomenon than when it is effected by ocular pressure, since, when the position of the prism and its distance from the object is known, it is not only possible to predicate its direction, but to estimate its.extent. It is an interesting fact that, in the case of a given distance, the prism produces or fails to produce a duplication of the image in proportion as the sight of the eye is more or less normal. This same remark applies to the vision of real objects in the waking state.

* Féré, *Soc. de Biologie*, October, 1881 ; *Progrès Médical*, December, 1881.

One of the present writers * has substituted other optical instruments for the prism, in order to verify and develop the former experiments. An opera-glass brings imaginary objects nearer, or makes them appear further off, just as if the objects were real. We begin by suggesting a given hallucination, either placed upon the wall of the laboratory or, which is better, on a screen covered with white; it may be a bird perched upon the wall, a lizard which runs up it, a flower, or a picture hung upon the wall. If the subject is made to look at the hallucinatory object through an opera-glass, it appears to be nearer or more distant, according to the end of the glass presented to his eye. It is well to guard against imposture, by not allowing the subject to see which end of the glass is in use. It is the simplest arrangement to have two cards of equal size, each pierced with two openings, and fastened with sealing wax to both ends of the opera-glass. This will prevent the subject from perceiving real objects on the field of the glass, which might, owing to the changes of dimension, serve as an index of its position. For the same reason, a plain and even surface should be chosen as a back ground for the hallucination.

It is interesting to observe that the opera-glass will not make the object appear more or less remote unless it has been adapted to the subject's sight. Thus W——, who is short sighted, could see nothing through a glass which had been adapted to C——'s long sight. Hence the subject must be requested to adapt the glasses to his sight before he is hypnotized. A great difference may be observed in the way in which subjects are affected by the experiment. C—— and D—— merely declare that the

* Binet, *L'Hallucination* (*Revue Philosophique*, April, May, 1884).

11

imaginary object is at one time near, at another far off; and this change of distance does not surprise them; only if an unclean beast is the object of suggestion and he is made to appear close to them, they utter a cry of terror. W——, who is much more intelligent, always evinces the most lively astonishment. When a bird perching on the branch of a tree is the object of suggestion, she cannot understand why it should be close to her at one moment and far from her at another. If we tell her that the bird is moving from place to place, flying nearer or further away, she rejects this explanation, and replies that the tree also seems to change its position. She finally concludes that her eyes are affected, so as to change the apparent distance of objects, which is a reasonable, if not a just conclusion. These reflections are all made during the state of somnambulism.

This experiment with the opera-glass may be modified by making use of hallucinatory portraits. The portrait of a given person may be made to appear on a square of white paper, and a series of experiments may be performed on this imaginary portrait, which are only a development of those with the opera-glass, since a final analysis must refer them all to an application of the laws of refraction. If a magnifying glass is placed before the imaginary portrait, the subject declares that it is enlarged, and if the lens is sloped, the portrait is distorted. If the sheet of paper is placed at a distance equal to twice the focal length of the lens, the portrait appears to be inverted. These experiments do not always succeed, but if under favourable conditions they are successful in only one instance, they must be accepted as genuine.

Again, a prism with three equal facets is passed over the white paper, and the subject is requested to look through it at the portrait, beginning with its upper part; he sees two heads instead of one, and these heads are enlarged in a direction corresponding with the orientation of the prism. Now, we must observe that the surface of the paper on which the prism is placed is perfectly white and smooth, so that no one ignorant of the properties of a prism could perceive that it doubled the image of the subjacent morsel of paper. If again the paper is applied to one facet of the prism, the subject sees only a single portrait, which seems to be folded in two. These appearances all conform to the reality; the subject would see the same series of modifications if there were really a picture on the paper. Under similar conditions a doubly refracting crystal gives two images, which are modified by revolving the crystal on its axis.

The last experiment of this nature is that with the microscope. The question arises whether, if a given preparation is supposed to be mounted on a microscopic slide, and the subject is made to examine it through the microscope, the hallucinatory image will appear to him to be magnified; whether this enlargement will be sufficient to reveal to him structural details which are invisible to the naked eye; whether, for instance, he will distinguish the corpuscles in the drop of blood suggested to be on the slide, or if he will see the stomata in a fragment of vegetable epidermis. The experiment is difficult, since most of the subjects who look at the microscopic slide fail to discover the imaginary preparation. Repeated attempts have convinced us that the microscope enlarges the hallucinatory image—that a spider's foot, for example, becomes enormous,—but we have not observed that hypnotic subjects discover details invisible to the naked eye.

It is much more easy to obtain the reflection of an imaginary object in a mirror. It may be suggested to the subject that an object is placed on a given point of the table, and if a mirror is placed behind that point the patient immediately sees two objects. If, for example, the appearance of one cat is suggested, a second is likewise seen, but the two objects are not always alike. On one occasion we gave to our subject the hallucination of a white cat, and the mirror caused another to appear, which was of a grey colour. The reflection of the imaginary object appeared to the subject to be just as real as the imaginary object of suggestion. Thus, when the mirror was in its place and the subject was told to look at the beautiful butterfly perched upon the table, she at once exclaimed that she saw two butterflies. When desired to catch them both, she made the gesture of seizing the one perched before the mirror, and fastened it with a pin to her bodice. This at least was done by D——, while C—— refused to be so cruel as to run a pin through the butterfly. She then tried to catch the second butterfly, of which she saw the reflection in the mirror, but as her hand came in contact with the glass, she was unable to reach the spot which the butterfly seemed to occupy. It was curious to observe W——'s behaviour at such a time. After knocking her hand against the glass several times in succession, she gave it up with indignation; disregarding our orders, she absolutely refused to repeat the attempt, saying, "I cannot do it, I cannot do it." It may easily be shown that the subject does not place the imaginary object on the surface of the mirror, but sees it *within* the mirror. In fact, if the mirror is advanced, withdrawn, or inclined, so that it could no longer reflect the supposed object, the double vision ceases.

These primary experiments are rude. They may be summarily explained by saying that the subject sees the mirror, and logically concludes that the object of suggestion must be reflected in it. We do not assert that this phenomenon of auto-suggestion is impossible, yet it seems to be excluded by the following experiments:—We recur to the hallucination of a portrait, which we have already employed, and shall have further occasion to employ. A prism of total reflection is placed upon the blank sheet on which there is an imaginary portrait. The resemblance of this prism to a mirror cannot warn the subject of what is to follow, and yet he never fails to see a second portrait, like the first one, when he looks at the hypothenuse facet of the prism. If the portrait is then placed opposite to a mirror, and it has been suggested that the profile is turned to the right, it appears in the mirror to be turned to the left. The reflected picture is therefore symmetrical with the hallucinatory image. If, without allowing the subject to see what is done, the paper is turned upside down, the reflected portrait also appears to be inverted, and, which is to be noted as in agreement with the laws of optics, the profile is turned to the right. We repeat that the imaginary portrait is turned to the right, and in the mirror it is turned to the left, but that if the paper is inverted, it appears to be turned to the right. Such combinations would not be invented, and a still more complex experiment may be performed. An inscription of several lines may be substituted for the portrait, and in the mirror it is inverted and runs from right to left. If the paper is turned upside down, so also is the reflection of the inscription, and at the same time it ceases to run from right to left. This experiment frequently, but not invariably, succeeds, and that in a way which excludes all suspicion of fraud. Yet few persons are aware that, while reflected writing must be read from right to left, this condition ceases when the reflected writing is inverted. Such difficulties do not exist for the hypnotic subject, who only sees and does not reason.

Since the imaginary object created by hallucination acts in all respects as if it were real, it may be asked whether that object is concealed by the interposition of a screen. This depends upon the subject, and the results are extremely varied. In the simplest case the hallucination is destroyed by the screen, and the subject declares that he has ceased to see anything. In the case of other subjects the screen has not this effect, the hallucination persists, without any change of place, and if the subject is ordered to seize the object of suggestion, his hand goes to the other side of the screen in search of it. In other subjects, again, the imaginary vision is not interrupted by an opaque body, but the object is transferred to that body.

We are unable to assign a cause for these variations, which may be noted in different subjects, and sometimes in the same subject, in the course of a series of experiments. We need only remark that peculiarities of the same kind occur in the vision of real objects during somnambulism. In some somnambulists the vision of real objects is not destroyed by the interposition of a screen, and it is destroyed in others. It must be clearly understood that these experiments were not concerned with the wonderful phenomena of vision through a thick bandage, of which so much was said in former times, and for the demonstration of which the Burdin prize was offered by the Academy of Medicine. We have not, strictly speaking, to do with vision *through* a screen, but with an hallucinatory vision which persists in spite of interposed screen, which is very different. The inconstancy of these phenomena, however, decided us not to make them the object of continued study.

The logical connection of the foregoing experiments will be readily seen. The first in date and importance is that of ocular pressure. This is a curious discovery, which must serve as a starting-point for a whole series of fresh researches, and it is the first instance of experiment on hallucinations, although it has been long neglected, and has only become fruitful in our own day. The prism experiment is merely a variant on that by Brewster; instead of the mechanical deviation of the eye produced by the finger, the prism causes the deviation of the luminous ray before it enters the eye, but the result of double vision is the same. The experiment with the opera-glass, again, may be regarded as a development of that with the prism, since both instruments are founded on the laws of refraction of light. Finally, the mirror is as closely connected with the preceding experiments as in physics the phenomena of the reflection of light are connected with the phenomena of refraction. All these new facts are logically derived from Brewster's experiment, which virtually includes them all, just as the properties of lines, angles, and surfaces virtually include the whole of geometry. It is only necessary to deduce and support each deduction by experimental research.

In order to give a satisfactory explanation of these experiments, we must decide between three hypotheses. 1. A suggestion is made; the subject is aware that a prism is placed before his eyes which has the property of duplicating objects, or an opera-glass which enlarges them, etc. But this first hypothesis must be rejected, since it is evident that the subject is ignorant of the complex properties of the lens, of the simple prism, the bi-refracting prism, and of the prism of total reflection, and although the subject may be acquainted with the other instruments, such as an opera-glass, care is taken to conceal them in the performance of the experiment. Therefore, unless we suppose that the experimenter has been incautious enough to announce the result beforehand, it must be considered certain that suggestion in this sense has had nothing to do with it. 2. The optical instruments employed have modified the real objects in the subject's field of vision, and these modifications have served as an index from which he could infer similar modifications in the imaginary object. Although this second explanation is better than the former one, it appears to us insufficient. It is opposed by numerous facts already cited—by the precise localization of the hallucination on a point which the experimenter is only able to find by the aid of elaborate measurements; by the recognition of an imaginary portrait on a blank card mixed with other cards which appear to us precisely the same; by the inversion of the imaginary portrait when the card itself is inverted, without the subject being aware of it, etc. We adopt a third hypothesis, which has already been indicated. 3. The hallucinatory image is associated with external and material marks, and the modifications produced by the optical instruments on these marks modify the hallucination in their turn. The following observations seem to confirm this theory.

We will first give an account of the experiments performed by Marie and Azoulay, on the duration of the perception of the imaginary object. These observers have shown that it takes a longer time to perceive an imaginary object than when the object is real.* "The apparatus consists of a white strip placed on Marey's cylinder; as the cylinder revolves, the strip passes before a spy-glass of somewhat small diameter, through which the subject looks. At the moment when the subject sees the white strip, he gives an electric signal. The precise moment at which the strip passes before the glass is known, so that in order to ascertain the time of personal reaction it is only necessary to measure the period which elapsed between that moment, and the moment when the signal was given.

"In the first series of experiments we estimated, in the case of an hysteric patient attended by Charcot, the time of reaction in the waking state, making use of a real white strip. This time was on an average 0·18″, and it was the same in the case of a normal individual. In the state of somnambulism the time was 0·20″, that is, it was increased by 0·02″. We then, instead of employing a real white strip, suggested to the hypnotized subject that there was a white strip on a certain part of the

* *Société de Biologie*, July 31, 1885.

blackened cylinder, although there was in fact absolutely nothing to distinguish this part of the cylinder. We told the subject to indicate the moment at which she saw the imaginary white strip. The time of reaction was 0·22″. We then awoke the subject and again estimated the time of reaction. In this case the average was 0·23″.

"It remained to be seen whether this average varied during the period of suggestion. On the following day, after the lapse of twenty-four hours, we found that the time of reaction was 1·02″, and after forty-eight hours it was 1·114″. We could not carry our researches further, since at the end of seventy-two hours the suggestion had vanished, and the subject could no longer see the strip on the cylinder.

"These are the two points on which we wish to insist: first, on the value of these experiments as a means of verification, since, as we were able to ascertain, simulation is absolutely impossible. However intent the subject may be, he cannot, either by sight or by the employment of some rhythm, succeed in producing such a tracing as those we have obtained, since in these tracings all the times of reaction agree almost perfectly. Therefore, the images furnished by suggestion may, as well as the real images, be checked by the graphic method. Secondly, the time of reaction increases enormously, but not in direct proportion with the duration of the suggestion.

	Seconds.
Immediately after suggestion it is	0·23′
Twenty-four hours after it is	1·02″
This is an increase of	0·79″
Forty-eight hours after it is	0 114″
This is an increase of	0·093″

"In a second series of experiments, the relative values of the time of personal reaction was the same, although each was slightly raised to the amount of 0·02″ or 0·03″."

The primary fact to be deduced from these experiments is that for the perception of a real object the time of reaction is 0·18″; for that of an imaginary object it is 0·23″. The probable reason for this difference is that in the vision of a real colour, there is only a single phenomenon—the sensation received by the eye. On the other hand, in the vision of an imaginary colour, fixed by suggestion on an external point, two things are involved—the vision of the point, and then the appearance of the imaginary colour on that point. This double phenomenon must occupy more time than a simple sensation. Moreover, as time goes on, the association between the point in question and the hallucinatory image is relaxed, until at last it disappears, since the moment comes when the point ceases to arouse any image in the subject's mind. Hence we understand why the duration of the time of reaction increases up to the moment when the perception ceases altogether. The graphic method has the advantage of seizing these progressive modifications in the duration of the imaginary perception, delicate phenomena which completely escape from simple observation. For this reason the experiments by Marie and Azoulay have a real psychological interest, teaching us how to measure the force of a mental association as it becomes weaker.

Londe, chief of the chemical works at the clinical establishment of the Salpêtrière, informed us of the following fact, which is a remarkable instance of protracted suggestion, confirming the ideas just set forth.

On one occasion when an hysterical patient was in a stato of somnambulism he approached her, and showed her a plate, representing a view in the Pyrenees, with asses ascending the hill, and said, "Look at your portrait. You are quite naked." The subject happened to see the plate when she awoke, and since she was furious at seeing herself represented in a state of nature, she jumped upon it, and broke it. Two photographs had, however, been taken from the plate, and these were carefully preserved. The patient's fury was excited whenever she saw them, since she always saw herself represented naked, and after two months had elapsed the hallucination still remained.

This extraordinarily long survival of the hallucination is easily explained by the theory of distinctive marks. In fact, photography presents to the subject an immense number of such distinctive marks, which become associated with the hallucinatory image and evoke it with invincible force by accumulating their effect. The most curious feature of this observation is that the subject did not see these distinctive marks, or, rather, it did not occur to her what they really were, since she must have seen them in order to project her hallucination. She failed to see, however, that their combination formed a view in the Pyrenees. The effort to convince the subject of her error was fruitless; she only saw her own portrait.

There is another remark to make on Londe's observation. This hallucination of the portrait existed in the case of all the proofs of the same photograph. It was first produced by the plate and then transferred to the copies printed off from it; the imaginary portraits equalled in number the printed copies. This multiplication of the hallucination by the multiplication of the distinctive marks somewhat resembles the phenomenon of the reflection in the mirror; at any rate, it clearly proves with what force the fictitious image was associated with the sight of the photograph, since the presentation of a new, but entirely similar copy was apt to suggest the same hallucination.

We repeat that if the hallucination of a portrait is created on a real photograph, the subject will discover the same portrait on a second copy of the photograph without the intervention of any fresh suggestion. This, which was at first an isolated experiment, led us to invent another which is logically derived from it. A blank card on which suggestion had fixed an imaginary portrait was photographed, and when we presented this photograph of the blank card to the subject, she instantly recognized the imaginary portrait. But it is evident that experiments of this kind are too delicate to be invariably successful. Just as experiments in physics sometimes miss fire, so it is with experiments in cerebral physiology. The one which succeeded had, however, a great demonstrative value, since it was the first. We made a second attempt, which completely failed, and we have made no further attempts.

This same theory of distinctive marks may explain some other points. It leads us to understand why hypnotic hallucination persists in many cases after the subject awakes, while the memory of what occurred during sleep is completely effaced, at least unless certain expedients are used to revive it. The contrast is striking. The subject is ordered to commit a murder, and on awaking, he neither remembers the order he has received, nor the act he is supposed to have performed. Yet if the hallucination of a bird has been given, it will be almost as vivid after awaking as it was during somnambulism, and for this we may ask the reason, since if the hallucination is an image, so also is memory. This is true, but the hallucination is an image with the addition of some external, distinctive mark, and this distinctive mark, which is still present, recalls the hallucinatory image by means of the association of ideas, like a knot tied in a pocket-handkerchief.

The existence of this distinctive mark establishes a natural transition between the hallucination and the illusion of hypnotism. The two phenomena are both produced by verbal suggestion; the only difference between them is that there is a substratum for the illusion which is wanting in the hallucination; in the case of every illusion there is a real object which is more or less transformed by the suggestion. It must be admitted that this is not an essential difference, since suggestion may transform the object in a thousand ways. A book may become a hat, a dog, or a person, and precisely the same appearances may be created without the aid of any object. For those who accept the theory of distinctive marks this difference between hallucination and illusion is completely effaced, and hypnotic illusion appears to be an hallucination for which the special suggestion has selected a distinctive mark which happens to be a real object. Hence some interesting consequences ensue; the hypnotic illusion is modified, just as an hallucination is modified in accordance with the real object to which it is applied.

We have seen that in the case of hallucination these

modifications consist in phenomena of deviation, duplication, etc., produced by optical instruments. In the case of illusion, since the distinctive mark is not a mark, but a real object, and often a person, it may be spontaneously modified, and this adds a fresh complication to the experiment. A gentleman named X—— was transformed by suggestion into a dog. The subject no longer saw X——, who ceased to exist for her, but she ascribed all his gestures and movements to the dog which had been suggested to her. Hence it followed that the illusion had not the fixed character habitual in hallucination, but constantly varied, and all the changes which took place in X—— reacted on the illusion. On one occasion we pointed out X—— to one of our subjects, saying, "Do you see that person? She is a nurse, carrying an infant." The hallucination persisted on awaking, and the subject looked at the nurse and child with feminine curiosity. Strangely enough, she watched X——'s gestures, and ascribed them to the nurse, so that the real and the imaginary were closely intermingled. When X—— raised his hands, she said angrily, "Wretched woman, is that the way to carry an infant? Do you wish to kill it?"

Hypnotic illusion leads us by a logical transition to the ordinary and physiological illusion of the senses, which occurs in so many different circumstances, and of which every one has had experience. The cause of these two illusions is not the same, since hypnotic illusion is produced by verbal suggestion, that is, from within, and the illusion of common life is generally produced by the imperfect perception of external objects, that is, from without. Yet it cannot be doubted, in spite of this difference in the formative process, that every kind of illusion is formed by the synthesis of two elements—the external element and the false image constructed by the mind and projected on the object. We may add the ordinary illusion may, like the other, be magnified by a lens, reflected by a mirror, etc., as some of our observations prove. In this case these optical modifications seem to be quite natural, since the false image is associated with an external object. It is, however, an interesting fact that a common law dominates the whole series of phenomena, hallucination, hypnotic illusion, and normal illusion.

Nor does the series end here. The normal illusion of the senses is directly connected with the external perception, that is, with the normal act by which we enter into relations with external and present objects. External perception is termed by Taine a true hallucination. Certainly this act is, like illusion, a synthesis of external sensations with internal images. The study of the mechanism of perception by one of the present writers leads to the conclusion that it presents on a small scale the phenomena which occur on a large scale in hypnotic hallucination—deviation, duplication, and enlargement of the mental images. Hallucination must, therefore, be a disease of external perception.[*]

The theory of distinctive marks may be extended to hallucinations which are given for some remote period; a singular fact which conflicts with our scientific experience, so that doubt has been thrown upon it.

* Binet (*La psychologie du raisonnement.* Paris, 1886) has set forth the principal psychological conclusions derived from these phenomena of hallucination.

Bernheim * said to S——, formerly a sergeant, when he was in a state of somnambulism, "On what day will you be at liberty during the first week of the month of October?" He replied, "On Wednesday." "Then, listen to me. Go to Dr. Liébault on the first Wednesday in October, and you will see the President of the Republic, who will give you a medal and a pension." He said that he would go, but remembered nothing of it when he awoke. On the 3rd of October, however, sixty-three days after the suggestion, S—— presented himself at Dr. Liébault's, at ten minutes to eleven. On entering he met and saluted F——, and then, without paying attention to any one, he went to the left side of the library, made a respectful salute, and uttered the words, "Your Excellency." As he spoke rather low, Dr. Liébault went up to him, and at that moment he extended his right hand, saying, "I thank your Excellency." Liébault asked to whom he was speaking, and he replied, "To the President of the Republic."

Beaunis informed the *Société de psychologie physiologique* (April, 1885) of another instance of post-hypnotic hallucination, which was realized six months after the suggestion. On the 14th of July, 1884, he told a hypnotized young woman that she would see him enter her room on the following 1st of January, and wish her a happy new year. And so, in fact, it was. On New Year's Day the young woman saw Beaunis (who was in reality in Paris) enter her room, wish her a happy new year, and disappear. Suggestions of dreams which would occur on some subsequent night have also been given. Our observations lead us to believe that hallucinations

* *De la Suggestion.* Paris, 1884.

given for some remote period really take effect. The length of the intervening period is less surprising than the fact that they are realized at the fixed hour.

It may be asked whether hypnotic subjects possess an abstract power of measuring time. We think it more probable that the occurrence of the hallucination at the moment assigned beforehand is produced by some external circumstance, and that if this circumstance, which acts as a stimulus, is removed, the latent existence of the hallucination may be indefinitely prolonged. It must be observed that in the experiments by Bernheim and Beaunis the day fixed for its fulfilment bore a distinctive mark. In the one case it was the first Wednesday of October, in the other the 1st of January. These dates perhaps served as the subject's distinctive mark; it was as if he had been told that the hallucination would occur when the experimenter clapped his hands, and the advent of the moment which had been named served as a kind of signal. This explanation must, however, be regarded as provisional, and the question, like so many others, remains open.

III.

We next propose to describe a series of hypnotic experiments which seem to throw some light on the still obscure problem of the physiology of hallucinations. In fact, the new phenomena with which we are about to acquaint our readers appear to show that hallucination is produced by an excitement of the sensory centres. If the conclusion is not new, it is, at any rate, interesting, and although it has often been set forth by physicians of the insane, no complete proof of the hypothesis has hitherto been given. Such a proof is furnished by the careful study of hypnotic phenomena.

We must, however, hasten to add that we are less anxious to develop a theory than to register a certain number of facts which are interesting in themselves inasmuch as they are facts. The conclusions which we draw from them with respect to the physiology of hallucinations are merely the bond which serves to connect very varied observations, and it is these alone which have any value. The regular observation of any phenomenon remains as a definite acquisition to science, whatever afterwards ensues, while the future of theories is uncertain.

1. *Achromatopsia.*—We will first consider achromatopsy, or colour-blindness. Paul Richer, in his *Études cliniques sur l'hystero-épilepsie,* was the first to show that in the case of most hysterical subjects it is impossible for their vision to accept hallucinations of colour. Since the eye has lost its chromatic sensitiveness, it cannot see the colours of an imaginary object.

As one of the present writers has shown, the same rule seems to apply to the spontaneous hallucinations of insanity. We observed in the St. Anne asylum, of which Dr. Magnan is in charge, an insane, hysterical patient who was constantly haunted by the image of a man dressed in red. The left side of this woman's body was affected by anæsthesia and achromatopsia, and when her right eye was closed, she continued to see the hallucination with her left eye, but instead of being dressed in red, the man appeared to be in grey, and enveloped in mist.*

* Binet, *L'Hallucination* (*Revue Philosophique*, April, 1884).

Since the fact is proved, it has still to be explained. It may be considered strange that a somnambulist subject is capable of receiving all sorts of hallucinations, however absurd they may be; that in the course of a few minutes she may be surprised by the appearance of a ball, a public *fête*, the overturning of a carriage, a tumultuous crowd, an insurrection, a conflict before barricades, succeeded by a quiet, moonlit night, revealing the bodies of the dead. The patient who sees all this may laugh, weep, be amazed, or utter cries of terror, in correspondence with the scenes unrolled before her; but when the attempt is made to display some coloured object to her, the experimenter's power is suddenly arrested, and the automaton, who is so docile in all other respects, obstinately asserts that she does not see the colours suggested to her. For instance, if the eye which remains open has lost the perception of violet, it is impossible for that colour to enter into any of her hallucinations, unless the other eye, which retains the sense of that colour, is opened. This is a striking fact, but it will cease to appear absurd when we bestow upon it sufficient reflection. It can be satisfactorily explained when we consider the seat of achromatopsia, and the probable seat of the hallucination.

It is now almost certain that hysterical achromatopsia results from a functional disturbance of the cerebral cortex, and not from any lesion of the retina, or of the media of visual perception. All that we know of the nervous disturbances of hysteria leads us to believe that they do not involve these media, and that we must regard achromatopsia as a functional disturbance of the cortical cells concerned with the perception of colour. This belief leads to the conclusion that if this functional disturbance is the same hindrance to the hallucination as to the perception of a given colour, it is probably because these two phenomena, perception and hallucination, employ the same class of nervous elements. In other words, hallucination occurs in the centres in which the impressions of the senses are received, and it results from an excitement of the sensory centres.

It may be objected that in the case of some hypnotized subjects achromatopsia is no hindrance to the suggestion of coloured hallucinations, but we find it easy to explain this deviation from the rule. The achromatopsy of hysterical subjects depends on hemi-anæsthesia, and there is nothing definite in this lesion. It is not so much a paralysis as a paresia, a slothfulness of the nervous elements. These elements no longer respond to the call of their normal excitant, the coloured ray, yet it is not surprising that they should react when approached from a different direction, by an excitement which proceeds from the auditory centres, and which is merely verbal suggestion.

2. *The phenomena of contrast.*—There is another fact which shows, still more clearly than the one just given, that hallucination and sensation have the same seat in the brain, namely, the property with which the hallucinatory image is endowed of producing the same effects of contrast as those produced by sensation. Parinaud, the head of the ophthalmological laboratory of nervous diseases at the Salpêtrière, has been good enough to send us the following paper, relating to hitherto unpublished experiments, which are of an extremely interesting character :—

"Hallucinations of colour may develop phenomena of chromatic contrast as readily as, and with even greater intensity, than the actual perception of colour.

"If, for instance, a piece of paper divided by a line is presented to a hypnotized subject, and it is suggested to her that one half is red, the sensation of the complementary colour, green, occurs on the other half. If, after awaking, the sensation of red remains, so also does the sensation of green.

"In order to understand the meaning of this fact, I must refer to the following experiment, relating to chromatic contrast, which I communicated to the *Société de Biologie* in July, 1882.

"A card which is half white and half green on one side, and wholly white on the other, is marked in the centre on both sides with a spot intended to fix the vision. For half a minute the eyes are fixed on the parti-coloured side, and then the card is turned and the eyes are fixed on the central spot of the white side. On the half which corresponds to the green half a red tint appears, which is merely the definitive after-image, and on the other half the complementary green tint is seen. The after-image of red has, therefore, developed by induction the sensation of green in the part of the retina which had only received the impression of white. This experiment, which may be varied in different ways, so as to establish the fact that we have to do with positive sensations, and not with any error of judgment, shows that every impression of colour leads to a more or less persistent modification of the nervous elements which produce the after-image and that this modification causes, in the parts not affected, a modification in the opposite direction which develops the complementary sensation, by a phenomenon analogous to that which occurs in a magnetized body.

"The image of hallucination acts like the after-image, and may likewise cause an induced sensation, so that it corresponds with a material modification of the nervous centres.

"In order that the experiment should succeed, it is necessary that the subject should retain the perception of the suggested colour, for we know that the perception of colours is frequently affected by hysterical amblyopia. If there is any blindness with respect to this colour, the suggested sensation is confused, and the induced sensation does not occur. When the subject is able to distinguish all the colours in the waking state, she can also distinguish their complementary colours. When only unable to distinguish certain colours, which is often the case, a singular result follows. Suppose that the subject sees red, and cannot see green, the hallucination of green cannot develop the induced sensation of red; yet the hallucination of red, which she sees, may develop the induced sensation of green, which she cannot see."

It clearly appears from these experiments that from the point of view of simultaneous contrast, the hallucinatory image acts precisely as a real sensation would do, whence it may be concluded that the two phenomena effect the vibration of the same keys of the cerebral instrument. They are, however, distinguished by the following difference. When a real sensation of colour is experienced, the sensation results from an excitement of the retina, and it reaches the centre of visual sensation by the paths of vision, by the optic nerve, the chiasma, the optic tracts, etc. The sensation of colour suggested by words, that is, the hallucinatory image, results from the excitement of the organ of hearing, and it is reflected in the centre of auditory sensation before it reaches the centre of vision. With the exception, however, of this difference in the process of excitement, we repeat that the hallucination and the real sensation appear to correspond with the same physiological process, since otherwise the same effects of chromatic contrast would not occur in both cases.

3. *Subjective Sensations.*—Parinaud's researches into the simultaneous contrast naturally led to the inquiry whether hallucinations produce subjective sensations, since these two orders of phenomena are closely allied.

Let it be remembered, for the sake of clearness, that the term of objective sensations is given to the perception of images which follows the impression on the sight of a luminous or bright object. The image which ensues is positive or negative, according to the conditions under which it is seen. In the positive image we have the representation of the object as it really is; its colour and the relatively luminous intensity of its parts are maintained. Conversely, in the negative image the light parts of the object appear to be dark, and the dark parts to be light; and its general colour is exchanged for the complementary colour.

The production of after-images is a normal phenomenon which constantly, but with varying intensity, accompanies the exercise of external vision. We have ascertained that hallucinatory vision is subject to

the same conditions; every hallucination of some persistence is succeeded, on its disappearance, by an after-
image, just as in the case of ordinary sensations which
affect the retina.*

This phenomenon was first observed, several years
ago, by the physiologist Gruithuisen. Observing what
occurred in his dreams, he states that "sometimes a
bright, fantastic image was succeeded by one of the same
form, but indistinct. Sometimes, again, after dreaming
of violet fluor spar, or burning coals, he perceived a
yellow patch on a blue ground." †

We had occasion to verify the exactness of this
observation in our treatment of hypnotic patients. The
somnambulist subject was requested to look attentively
at a square of white paper, with a black spot in the centre,
designed to fix her vision. At the same time the suggestion was made that the paper was of a red or green
colour. A second square of paper, likewise marked in
the centre with a black spot, was then produced, and as
soon as the subject had fixed her eyes on the spot, she
exclaimed that the spot was surrounded by a coloured
square, and the colour indicated was complementary to
that which had been made to appear by means of suggestion. This complementary colour is the negative image
left by the hallucinatory colour; it lasts but a short
while, it is effaced, is lost, or dies, as the subject says,
and it resembles in all respects a normal, negative image.

This experiment was repeated by Charcot before a
numerous assembly, during one of his lectures on aphasia.

* This experiment was performed for the first time by Richer and the
present writers in June, 1884.

† Quoted by Burdach, in his *Traité de physiologie*, vol. v. p. 206.

12

That eminent professor demonstrated that, in order to ensure success, care must be taken to define the nature of the suggested colour. For instance, if only the colour *red* is suggested, the subject may either see the shade of red, of which green is the complementary colour, or the orange-red, of which blue is complementary. These contradictory results are impossible when the colour which the subject is intended to see is made clear by a comparison. It may be remarked in passing, that this experiment is a peremptory reply to those who still believe in the existence of a general simulation. It would be unreasonable to maintain that an hysterical woman, who scarcely knows how to read or write, has the theory of complementary colours at her fingers' ends. Our subjects have always answered correctly, and, which is more important, the correct answer has been given when the experiment was performed for the first time.*

It must be remembered that analogous phenomena occur in the mental vision of normal individuals. The persistent idea of a brilliant colour develops an after-image of the complementary colour, just as a real sensation does.† If we close our eyes and fix our minds for a long while on an image of some vivid colour, and then open our eyes to look at a white surface, we shall, for a brief space, see the object of our imagination, but of the complementary colour. One of the present writers successfully repeated this experiment, which is difficult and demands from the subject a great power of visualiza-

---

* An interesting fact was displayed in the case of one of our subjects. She had lost the perception of violet in both eyes; violet looked like black. When the hallucination of yellow was given, the after-image was black, instead of violet, the complementary of yellow.

† Wundt, quoted by Ribot, *Maladies de la mémoire*, p. 11.

tion. He was able to picture to himself the idea of red, so intensely that at the end of a few minutes he was able to see a green patch upon the white paper ; but, strangely enough, repeated efforts were required before he was able to associate an outline with the colour, so as to reproduce under the form of a subjective image the idea of a coloured cross or circle.

These facts show the close connection which unites sensation, hallucination, and memory. These three phenomena are evidently based on the same physiological operation, and are effected in the same region of the nervous centres. Thus, whether it is a real impression of the colour red, whether the colour is pictured by the memory, or again, whether it is seen by an hallucination, it is always the same cell which vibrates.

4. *The Mixture of Imaginary Colours.*—Since it is interesting to develop an experiment, so as to consider a fact in every aspect, we sought to discover what would result from the mixture of imaginary colours. We wished to know whether an hypnotic subject could make white out of the suggested colours of combined red and green. The process which, after many attempts, we found to be the most convenient, does not involve much preparation. Two squares of coloured cardboard are placed on a table at a little distance from each other, and a piece of glass is held before the eye at such an angle as to admit a direct view of one of the squares, and at the same time a reflected image of the second square ; in this way it is easy to place one image over the other, and thus to mix their colours. The result may be varied in manifold ways by employing differently coloured squares. After this arrangement is made, a series of blank cards are

shown to the subject, and it is suggested to him that they are coloured. Care is taken each time to define the suggested colour by showing to the subject, by way of pattern, one of the coloured cards which were used for the previous experiment. In this way the imaginary colours of the white cards are absolutely the same as the real colours on the other cards.

The subject may then, with a piece of glass and his collection of cards coloured by suggestion, effect the same mixture of colours as the experimenter, who can on each occasion verify the exactitude of the result by the combination of the real colours. Under these rigorous conditions, which leave no scope for erroneous suggestions, the imaginary colours give the resultant shades which are in conformity with optical laws. Hence we may conclude that the hallucination of a colour is a suggested sensation which occupies the same region of the cerebrum as a real sensation.

5. *Phœnomena observed with reference to the Eye.*— We now come to a consecutive series of clinical observations and of experiments which furnish a valuable argument in proof of our thesis, and which are perhaps the most decisive of all.

When a lesion of the brain produces sensory disturbances in the integuments of the eye, visual disturbances also occur, such as achromatopsia, or concentric or lateral contraction of the field of vision. This has been shown by numerous observations.* This singular relation between the general sensitiveness of the eye and its

* Féré, *Des troubles fonctionnels de la vision par lésions cérebrales,* pp. 152, 153 (1882); *Notes sur l'anesthésie hystérique (Soc. de Biologie,* October 29 and November 26, 1881; July 24, 1886).

special sensitiveness is particularly apparent in the hemi-anæsthesia of hysterical patients. In fact, in these cases, the insensibility of one-half of the body not only extends to the skin and mucous membrane, but also to the other organs of the senses: sight, smell, and hearing are likewise affected on the same side; in a word, there is, as a rule, a sensitivo-sensorial hemi-anæsthesia. Under these conditions the general sensitiveness of the eye, the sensitiveness of the conjunctiva and of the cornea, is always in correspondence with the special sensitiveness of that organ. Thus the hemi-anæsthetic hysterical patients whom we have observed, and in whom there was neither contraction of the field of vision, nor achromatopsia, retained the sensitiveness of the conjunctiva. Those who had lost the power of seeing one or more colours, or whose field of vision was more or less relatively contracted, did not only experience anæsthesia of the conjunctiva, but also of the cornea. In this latter case, if while the subject is looking intently at any object, the conjunctiva and the cornea are touched by a strip of paper, the eye and eyelids do not move as long as the foreign substance does not come in front of the pupil. The reflex movement of the eye and eyelid which occurs in the latter case is exclusively produced by the excitement of the retina, which has lost the perception of colour, but still distinguishes light and shade.

In those who are simply hemi-anæsthetic, or who are totally anæsthetic with a predominance on one side, magnetization, static electrization, etc., will effect a transfer of the anæsthesia, so as to furnish a counter-proof, affording constant results.

This relation between cutaneous and sensorial insensibility does not only exist when the anæsthesia extends to one-half of the body, but when it is more restricted. When statical electrization has destroyed the hysterical anæsthesia, after an interval which varies in different subjects, insensibility reappears in a localized region, which does not correspond with the distribution of the nerves. In the case of one of our patients, insensibility first returned to a limited zone round the eye, which included the cornea and the conjunctiva, and visual anæsthesia was reproduced, simultaneously with the limited anæsthesia of the skin.*

Another proof of the relation which exists between the special sensitiveness of the eye, and the sensitiveness of the conjunctiva, may be found in the observations we made on three hysterical and hypnotic patients at the Salpêtrière. Two phases of catalepsy may be distinguished, as far as the eye is concerned : First, in profound catalepsy, such as is produced by a sudden noise, the eyes remain fixed, with no movement of the eyelids. In this state it is possible to touch the conjunctiva without producing any reflex action. Secondly, when an object is moved to and fro before the subject's eyes, so as to fix his gaze on the moving object. If in this case the conjunctiva is touched, there is an immediate reaction of the eyelids, as in a healthy subject, although the insensibility of the rest of the body is maintained. The experiment may be repeated at pleasure by again throwing the subject into a state of profound catalepsy, and the result is always the same;

* What we have said of the eye applies also to the other senses. For further details, into which we do not enter here, see Féré's work, quoted above.

as soon as the fixed gaze ceases, the sensitiveness of the
conjunctiva returns. The object which is moved to and
fro excites the special sensitiveness of the eye, just as
under other circumstances a strong local excitement
brings back the sensitiveness of the skin, and together
with the function of sight the sensitiveness of the external
membrane of the eye returns.

These facts appear to indicate that in some indeter-
minate region of the brain there are sensory centres
common to the organs of the senses and to the integuments by which they are covered.[*]

This long preamble brings us to the observations in
which we are more immediately interested, which regard
the physiology of hallucinations. One of the present
writers has ascertained that when a cataleptic subject
receives a visual hallucination, the general sensitiveness
of the eye is often profoundly modified. We have just
seen that in the cataleptic state the conjunctiva and the
cornea, with the exception of the region of the pupil, are
generally insensitive. In the case of the subject P——,
as soon as a visual hallucination was developed, the
sensitiveness of the external membranes of the eye
returned; the membranes could not be touched by
any foreign body without producing reflex action of the
eyelids.[†] An hallucination arouses the general sensitiveness of the eye, just as it is aroused by waving a
real object before the subject's eyes. This is surely a
proof that visual hallucination excites the visual centres.

A second experiment displays the same fact under

[*] Féré, *Troubles fonctionnels de la vision*, pp. 149, 151, 158.

[†] Féré, *Les hypnotiques hystériques, comme sujets d'expérience en
médicine mentale*, etc. (*Arch. de Neurologia*, p. 122, vol. vi. 1883).

another form. In M——'s case the visual hallucination usually persisted for three or four minutes after awaking. As soon as this subject awoke, she complained of pain in the eyes and constantly rubbed them, only ceasing to do so when the hallucination disappeared. We saw this behaviour repeated more than forty times without attaching any importance to it, so true is it that we only see what we expect to see. And yet the phenomenon is curious; the hyperæsthesia, or rather the dysæsthesia, of the integuments of the eye is produced by the visual hallucination, since it lasts for the same period and disappears at the same time. If an hallucination brings about this modification of the cutaneous sensitiveness of the eye, it is probable that it excites the special sensitiveness of that organ, in other words, the centre of vision.*

We observed another form of the same phenomenon in the subject X——. We suggested to her the hallucination of a bird perched on her finger, suggesting at the same time that she only saw the bird with her right eye. The hallucination persisted after awaking; the subject fondled the bird without being aware that she only saw it with one eye, for both eyes were open, nor did it occur to her to close one of them. After a while she complained of pain in the right eye, saying that she felt as if sand had got into it, and she only put her hand to this eye. It should be observed that persons affected by conjunctivis complain of the same sensation of sand in the eye. The localization of the pain in the right eye proves that the dysæsthesia depended on the hallucination.

* In this patient an hallucination of hearing produced local pain in the auditory meatus.

Each of these facts, taken alone, is insignificant, but they are in logical agreement and connection, and appear to prove that visual hallucination has its seat in the visual centre.

After studying the influence of hallucination on the organs of the senses, and on the eye in particular, we must mention the observation made by one of the present writers* on the state of the pupil in subjects of hallucination. He observed in the first place that in the hallucinations which accompany the third period of a strong hysterical attack, the diameter of the pupil varies with the assumed distance of the hallucinatory object. This interesting fact also occurs in the hallucinations produced by hypnotism: "This was what we observed in the case of two hysterical patients with whom it was possible to hold oral communication during catalepsy. When we desired them to look at a bird perched on a steeple, or flying high in the air, the pupil was gradually dilated until its normal diameter was almost doubled. When we caused the bird to fly down, the pupil gradually contracted, and the same phenomenon was reproduced as often as the idea of any moving object was evoked.

"The modifications of the pupil produced in this manner in a cataleptic subject, who continued to display all the phenomena peculiar to catalepsy, show that the fictitious object of hallucination is seen just as if it actually existed, and its supposed movements produces efforts of accommodation which are governed by the same

* Féré, Note on some phenomena of the eye observed in h stero-epileptic subjects, either during the attack or at other times, in the *Société de Biologie*, October, November, December, 1881.

laws as if it were a real object. We, therefore, have to do with a true hallucination, and not with any imposture."

IV.

We have now to observe how hallucination is affected by æsthesiogens.

We have had frequent occasion to speak of æsthesiogenic agents. The term is applied to certain agents which, according to Burq, whose observations have been confirmed and extended by other scientific men, have the property of acting on the sensibility and motor power of a certain category of subjects. The magnet is the æsthesiogen to which we have most frequently had recourse. There is nothing mysterious about this agent; compared by physics to a solenoid, it acts like a faint electric current on the nervous system, and produces a continuous peripheral excitement. Its mode of action has, moreover, been clearly established by one of the present writers.[*]

We need not in this place prove the reality of æsthesiogenic influence, in order to reply to those who only see in these agents the effects of suggestion and of expectant attention, since we have already had occasion to explain this point. It only remains to show that in the following experiments we took sufficient care to eliminate suggestion and expectant attention. These were the points on which we insisted: 1. Since these researches were new to us, we were in many cases unable to foresee what would occur, and especially with respect to the polarization of emotions, so that suggestion

* Féré, *Bull. Société de Biologie,* p. 590, 1885.

on our part was impossible. 2. We repeated the experiments on absolutely fresh subjects, and obtained the same results. 3. The same effect was produced when the magnet was concealed under a cloth. 4 This was also the case when the magnet was made invisible by suggestion. 5. We made use of a wooden magnet, and nothing occurred, although if there had been any results they could not have afforded a counter-proof, since they might have been explained by the recollection of a previous peripheral excitement. 6. The experiments made under somnambulism were in logical connection with those made under lethargy and catalepsy, although in these two latter states we have found our subjects incapable of receiving any complex suggestion. We think that under these conditions the results we have obtained must be considered due to æsthesiogens, and not to unconscious suggestion.

Many observers may attempt to verify our experiments, and if they fail they will declare them to be false, or to have been produced by suggestion. Let them remember Claude Bernard's remark, that for the most part a negative experiment only proves that it has been imperfectly understood. It is clear that æsthesiogens only act on a certain class of subjects ; this was admitted at the outset by all observers who have studied this question. Since our experiments with æsthesiogens are only the logical development of the experiments in metallo-therapeutics made by Burq and his successors, they are clearly subject to the same conditions. Our researches will not be invalidated by showing that they failed with the first subject who was presented to them; such an argument would be irrational. We do not disbelieve the phenomenon of neuro-muscular hyper-excitability, because it cannot be produced in a healthy subject, not affected by hysteria. Our present and future opponents are recommended to perform their experiments exclusively on the hysterical patients who display the features of profound hypnotism, and in those whose sensitiveness and muscular strength are modified by the application of magnetism.

We have observed that in the case of some subjects affected by profound hypnotism, unilateral hallucination may be transferred by the magnet, like a contracture or an hysterical paralysis.*

Contrary to what occurs in contractures, the transferred hallucination of vision is not symmetrical with the initial hallucination. It is suggested to the subject that he sees a portrait in profile on a card, and that this profile is turned to the right; it is added that he only sees this profile with the right eye, not with the left. By applying a magnet the hallucination is transferred from the right to the left eye; if the subject is then asked to which side the profile is turned, he says, as before, that it is turned to the right, although symmetry demands that it should be turned to the left.

During the transfer, the subject spontaneously complains of pain in the head, shooting from side to side. This pain is not diffused but local, and its seat is noteworthy. The cranio-cerebral topography established by one of the present writers † enables us to show that the point where pain is confidently indicated by the subject

---

* Binet et Féré, *Le transfert p ychique* (*Revue Philosophique*, January, 1885).

† Féré, *Anatomie médicale du système nerveux*, p. 95. Paris, 1886.

coincides, in the case of certain forms of hallucinations, with the sensory centres of the cerebral cortex, just as they have been established by the physiological and anatomical researches of late years (Fig. 9). This is

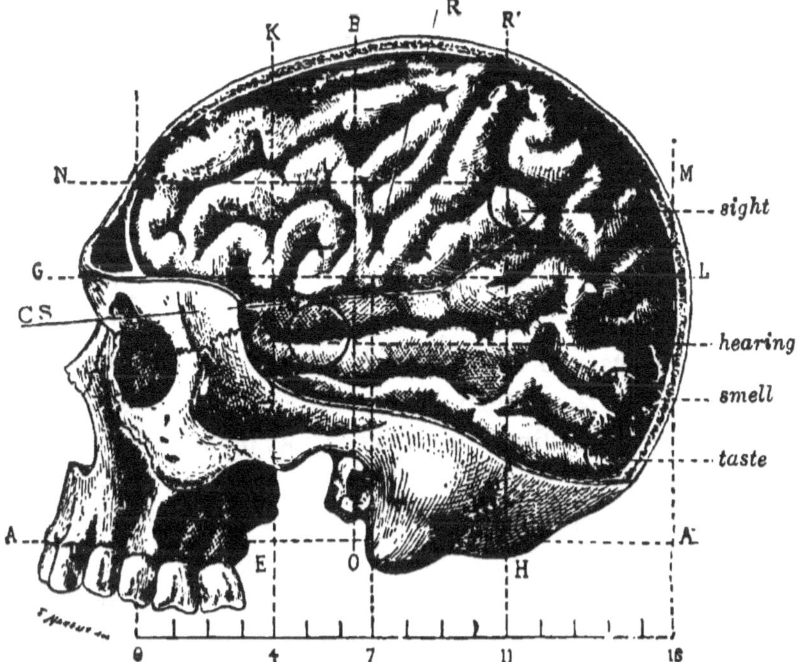

Fig. 9.—*Cranio-cerebral topography.*—B, Bregma ; C, Point corresponding with the outer extremity of the coronal suture; L, Lambda, corresponding with the parieto-occipital fissure; C S, Fissure of Sylvius; R R'. Sulcus of Rolando; R. Its anterior extremity, about three centimeters behind the outer extremity of the coronal suture; R', Its posterior extremity, forty-five millimeters behind the bregma; A A, Condylo-alveolar plane; B O, Auriculo-bregmatic plane; G L, Plane passing through the minimum frontal transverse diameter and the lambda; K E, Section passing between the two folds of the third frontal convolution at the point C, which corresponds with the external extremity of the coronal suture, and just touching the head of the caudate nucleus; R' H, Section passing through the posterior extremity of the sulcus of Rolando, and behind the posterior border of the optic thalamus; N M, Horizontal plane passing over the upper surface of the corpus callosum, and below the grey nuclei.

especially the case with the most important hallucinations, those of sight and hearing. Thus, in the transfer of an hallucination of vision, the point is a little behind and above the pinna of the ear, corresponding with the region of which the destruction causes blindness and hemianopia; it is, therefore, the posterior part of the lower parietal lobule.

In the transfer of an hallucination of hearing, the pain is seated in the centre of the space included between the anterior part of the pinna of the ear and the external, angular process of the frontal bone. The pain almost corresponds with the centre of the temporo-sphenoidal lobe, and approximatively with the region of which the destruction causes deafness. For the sense of taste, the point is above the external occipital crest, two centimeters from the median line. For the sense of smell, it is one centimeter above that line. These two latter localizations are not in agreement with the results of anatomical and clinical researches, and demand revision.

It may be asked how this coincidence should be interpreted: whether it proves that the physiological process in correspondence with the hallucination is seated in the sensory centres of the cerebral cortex, behind the motor zone; or if we are only to regard it as one of the reflex acts, termed in physiology, an echo of pain. We cannot decide this question, since it is only certain that in the case of some subjects there is a special relation between some points of the external covering of the head, and certain nervous centres of which the exact locality is still undefined. On this account the seat of the pain of transference must be estimated as an objective sign.

In another experiment on the same subjects, we obtained a demonstration of the same relation between certain points of the hairy scalp and certain sensory functions. This was in experiments on partial somnambulism.* If the subject is thrown into a state of total catalepsy, and those points of the scalp which become painful during the transfer are then mechanically excited by the finger or by some other blunt object, curious results are produced. On exciting that point of the scalp which corresponds with the centre of vision, both the subject's eyes are affected by somnambulism; they lose their cataleptic fixity, and follow the movements of the finger. If the point corresponding with the auditory centre is excited in a similar way, somnambulism affects the organ of hearing, and the subject who, up to that movement is completely insensible to the voice, hears the orders addressed to him and attempts to execute them, so far as his limbs, which are still cataleptic, allow.

We have seen what effect is exerted by the æsthesiogen on unilateral hallucinations; it displaces, and subjects them to a series of oscillations. When the hallucination is bilateral, the result is different; it is not a transfer, but what we have termed a polarization.† Of this we will give some instances.

The usual hallucination of a bird perched on her finger was given to a somnambulist subject. While she was caressing the imaginary bird, she was awakened, and a magnet was brought close to her head. After the lapse of a few minutes, she suddenly paused, raised her eyes

* Féré and Binet, *Le somnambulisme partiel* (*Soc. de Biol.*, 1884).

† Féré and Binet, *La polarization psychique* (*Revue Philosophique*, April, 1885).

and looked about her in astonishment. The bird which she supposed to be on her finger had disappeared. She looked about the room, and finally discovered it, since we heard her say, " So you leave me thus ! " The bird presently disappeared again, and once more reappeared. The subject complained from time to time of pain in the head at the point already described by us as corresponding with the centre of vision.

The magnet exerts the same suspensive effect on a real perception. For instance, after awaking one of our subjects, we showed her a Chinese gong and the pad with which it was sounded. The sight of the instrument alarmed the subject, and as soon as it was struck, she fell into a state of catalepsy. After this preparatory experiment, she was awakened and requested to look attentively at the gong, and meanwhile a small magnet was brought close to her head. In the course of a minute she asserted that she could not see the instrument, and that it had completely disappeared, and when the gong was sounded, more loudly than before, she did not fall into a state of catalepsy, but only looked about her with an air of some surprise. Hence it appears that the magnet in some sense paralysed the vision of the gong; the perception of this object was replaced by a corresponding anæsthesia, so that the noise of the gong no longer produced catalepsy.

We have also ascertained that the magnet destroys a suggested memory, just as it destroys both real and imaginary vision. This analogous effect is intelligible, since all these phenomena have a common basis. Memory is an image, and so also is hallucination, and an image is only a faint copy of an anterior sensation.

Memory, hallucination, and true perception are distinguished by the secondary states of consciousness which accompany the suggestion of the image. In the case of memory, this state consists in the reasoning which localizes the image in the past. In hallucination and true perception these states consist in reasoning which localizes the image in the external world. But these localizations in space and time are secondary, accessory, superadded acts.

One experiment on polarization clearly shows the connecting link between these three phenomena. When one of our subjects was in the waking state, we spoke to her of the gong, begging her to describe its form, colour, size, and use. She repeatedly told us that she saw it distinctly in her mind. When her attention had been firmly fixed on the idea of this object, we applied the magnet, and in the course of a minute she had a difficulty in picturing the gong to herself, and ended in being unable to understand us when we spoke of it. We then took the gong off a table which stood near, and offered it to the subject, who did not see it. Even when it was sounded with considerable force, she only gave a slight start. But after waiting a few seconds, a consecutive oscillation took place; her recollection of the gong returned, together with the vision of the instrument, and then a slight stroke upon the gong sufficed to throw the subject into a state of catalepsy.

Thus we see that the suppression, or rather the paralysis of memory, produced by the application of the magnet, induces a corresponding paralysis of the perception of the object. Since the subject was unable to picture the gong to herself, she was also unable to see it when it was presented to her.

The foregoing account describes the action of the magnet on a sensation, an hallucination, a memory, either by suppressing them, or by substituting a corresponding

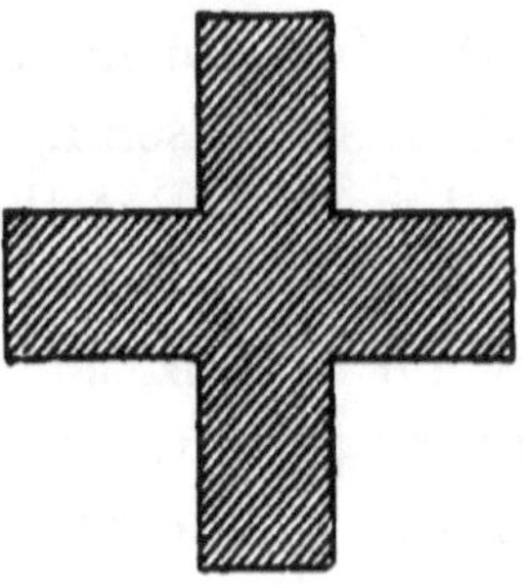

Fig. 10.—Red cross.

paralysis. There is an additional element in polarization, the production of a complementary phenomenon. This

Fig. 11.—Rose-coloured cross with green rays between its arms.

is shown by the following experiment, to which we must limit our description. We have seen that after looking intently at a red cross, and then on a white

surface, a green cross appears as the consecutive sensation; in the subjective image the primitive colour is
replaced by its complementary, but the form of the cross
still persists. The same thing occurs when the hallucination of a red cross is evoked, or when the same coloured
figure is presented to the imagination with sufficient
intensity.

Fig. 12.—The cross has disappeared, leaving a space. The green rays
have become elongated and of a darker shade.

If we inform one of our subjects, W—— or C——,
while in the waking state, that the cross we have just
drawn on a piece of white paper is coloured red, and if
we then request him to consider this cross attentively
while a magnet is, without his knowledge, placed behind
his head, we have the following result: the subject sees
green rays appear between the arms of the cross, these
rays gradually become elongated and at the same time

the original red tint of the cross changes to pink. For an instant the cross appears to be green, and then all colour disappears from the original figure, and the subject perceives a blank cross, a space surrounded by the persistent green rays (Fig. 12, p. 271). If at this moment a red cross is placed over the outlined figure, the subject is unable to see it. The magnet produces analogous effects on the memory of coloured objects.

It would be interesting to apply similar experiments to the senses of taste, smell, and hearing, in order to ascertain whether in the case of these sensory organs it is possible to establish a theory of complementary sensations, comparable to those of vision. We have merely had occasion to observe that under the influence of the magnet a suggested impression of heat is replaced by an impression of cold, which causes shivering.

V.

In the case of some subjects, the hallucination begins and ends during somnambulism. In others it is more permanent, and persists during the waking state. It will be easily understood that the duration of post-hypnotic hallucination is very variable, since it depends on many circumstances.

In the case of subjects whose hypnotic hallucinations continue in the waking state, it is interesting to inquire what the hallucination becomes in its new environment. It might at first appear that the subject would on awaking correct her hallucination and in some sense expel it from her intelligence, but this is by no means the case. In subjects affected by profound hypnotism the hallucination which persists in the waking state is accompanied by blind confidence. It is useless to tell the subject that she is the victim of an illusion, and that the portrait which she thinks she sees is an imaginary vision. She regards such language as a mockery, and if it is repeated the subject becomes uneasy and assumes a distracted air, and indeed on one occasion an hysterical attack appeared to be imminent. These facts seem to show that a conviction of the reality of the hallucination is an essential part of the phenomenon; the hallucination does not merely consist in the external projection of a sensible image, but in the condition of mind which accompanies the projection of this image.

On one occasion we informed our subject before hypnotizing her, that we were going to suggest an hallucination, and we agreed that she should, on awaking, make every effort to correct the hallucination and regard it as false. After hypnotizing her, we gave her the suggestion that a gold ten-franc piece, bearing the effigy of Napoleon III., was lying on the table. When she awoke, she still saw this coin. We said to the subject, "You know what we agreed upon. We gave you an hallucination, and the gold coin is not really there." She looked at us with a stupefied air, or I might almost say in a stupor, our words seemed to her so amazing. The mere idea that it was possible to doubt the existence of a piece of money which she saw and touched seemed to disturb her intelligence. But she soon recovered herself, and positively declared that she saw the coin, that it was a real coin, and that we were laughing at her in asserting the contrary. It was impossible to infuse the slightest doubt into her mind. The hallucination might be destroyed by suggestion, but as long as it remained, the subject believed it with all her might.

On the other hand, many writers have declared that it was only necessary to tell the subject that the hallucination was suggested to destroy his belief in the reality of the vision. We have only met with this submission in our subjects when the hallucination was fading away, and had lost its intensity.

We have still to say how the hallucination which has been produced may be destroyed. It is often very important to leave no trace of a sensory disturbance which may have dangerous consequences for those who are brought in contact with the patient. Any subject of hallucination is as dangerous as some explosive substance. In many cases the hallucination is spontaneously effaced, and this mode of disappearance has been well described by some subjects. The imaginary object loses its distinctness of outline, it becomes transparent, and ceases to conceal the real objects before which it is placed, and it finally seems to melt into air (Richer). In other cases, the hallucination disappears during the waking state, after an interval which varies with the subject. Some are in despair when the imaginary object disappears. One subject, to whom Bernheim had given imaginary rings, bracelets, and fans, implored him to leave her in possession of these gifts, since experience had taught her their fugitive nature. Others try to find some mode of accounting for this strange disappearance. X——, who, at the end of a few days, saw the portraits which had been suggested to him fade away, and the cards become blank, explained the fact by saying that the photographs had been badly focussed.

The simplest mode of destroying the hallucination is to assure the subject that he has seen, heard, and felt nothing. Sometimes he resists this intimation. The magnet may also be employed, if the subject is sensitive to this æsthesiogen; we have already seen that the magnet rapidly destroys a bilateral hallucination. For the most part, the hallucination and all recollection of it disappears together; such amnesia may even occur when the hallucination has been produced in the waking state, and this is a valuable note by which we may ascertain the sincerity of the experiment.

This is the place for mentioning some curious phenomena. A somnambulist was shown a real scent bottle, which was on the table, and it was then removed, and she was told that it was still there. On awaking she saw the imaginary scent bottle, and was unable to see, smell, or in any way perceive the real one. It was placed in her hand, passed over her face, struck with a key, and she was not the least aware of it. The perception of the real object was completely paralysed by the imaginary vision of the same object.

Here is another example. The hallucination was given to X—— that one of the present writers attended a ball which is annually given at the Salpê-trière; she saw him distinctly and spoke to him several times during the ball. He came to the Salpêtrière on duty on the following day, in reality this time, and not in hallucination. X—— saw, but did not recognize him, taking him for a stranger. It was necessary to hypnotize her in order to restore the perception of his person.

An hallucination may also be destroyed by a simple physical excitement. It is suggested to a somnambulist that she hears a letter repeated, *L*, for example. On awaking she continues to hear the same sound. On opening her mouth it is ascertained that the movements of her tongue coincide with each mental act of hearing. If this movement is checked by an energetic pressure, the hallucination disappears. It also disappears if the subject protrudes her tongue, and keeps it in this constrained attitude, or again if a contracture is produced.

We see, therefore, that an hallucination may be destroyed by three different processes, by suggestion, by physical excitement, and by the magnet. It is probable that the two latter agents really act by suggestion.

Hallucination, of which we have now described the chief characteristics, stands at the beginning of a series of much more complex and more obscure phenomena; it may therefore serve as the preparation and introduction to the study of these higher phenomena, among which we may mention the conceptions of delirium.

CHAPTER X.

MOTOR suggestions present us with a series of experiments which start from a simple, natural, and even fairly intelligible phenomenon, the suggestion of a movement, and culminate in complex phenomena which it is most difficult to explain, the suggestion of acts. Acts do not consist merely of movements, but of sensations, perceptions, reasoning, reflection, and will. The act may be said to be the resultant in which all the intellectual, moral, and motor functions of the individual converge.

The simplest suggestions of movement belong to the cataleptic phase. We have seen that harmony is the chief feature of the attitudes artificially impressed upon the subject.

The expressive movements given by the experimenter to different parts of the body are always immediately reflected in the countenance, which thus completes the expression. Braid clearly saw this influence of the gesture on the countenance, and we regard this fact as one of the finest and most wonderful results of hypnotic experiments. It affords to psychology a valuable source of information in the expressive mechanism of the emotions, and it furnishes the artist with a motionless model.

13

representing with striking fidelity all the sentiments of which man is capable. It has been said that the sculptors of antiquity took women in a state of catalepsy for their models, and there is nothing improbable in the fact.

An infinite number of expressive attitudes may be given to the subject, who may be caused to express ecstasy, prayer, grief, suffering, disdain, anger, and fear. If the extended hand is approached to the mouth, as if in the act of sending a kiss, the mouth smiles. If the fists are clenched, the brows contract and the face expresses anger. This reaction of the gesture on the countenance is not only seen in catalepsy but in other states, as, for instance, in somnambulism, and even, although in a minor degree, in the waking state. But the maximum of intensity only occurs in cataleptic subjects, on account of the complete automatism which characterizes this state. The slightest change in the attitude of the limbs produces a corresponding modification of the expression. When the open hand is carried to the lips, there is a smile; when the fist is clenched, there is an expression of anger. Moreover, this reaction of the expression takes place, with whatever rapidity the change in the attitude of the limbs may be effected. If the subject's open hand is taken and moved swiftly to the mouth and again withdrawn, a formal smile is seen on the lips as the hand approaches, which passes away as soon as it is withdrawn. Again, the influence of the gesture on the countenance may be rendered unilateral; when the left hand is clenched, a frown is seen on the left side only, and if at the same time the right hand is approached to the mouth, there is a smile on the right side of the face. Each side of the face thus expresses a different emotion.

It occurred to Charcot and Richer to modify the gesture by acting on the countenance. By means of localized faradization they developed a given emotion on the features, and the body at once assumed an attitude in correspondence with that emotion. When once it has been produced, the emotion impressed upon the features does not become effaced, and the position of the limbs is likewise persistent. It is possible, by graduating the force of the current, to cause the subject to express different degrees of the same emotion. We have seen that in a state of lethargy, all the muscles of the face may be contracted separately, by pressing the finger on their motor points. The experiment of which we are now speaking is of a somewhat different character, and of greater importance in the study of the play of countenance. The excitement is no longer, as in lethargy, localized in the muscle which is touched; it is communicated to the other facial muscles, which must also be brought into play in order to produce the desired expression.

It has often been asked what is occurring in the mind of the cataleptic subject, when he is placed in an emotional attitude. There is a curious contrast between his statuesque immobility and the tragic expression of his countenance; in one sense he appears to see and hear nothing, and in another he displays intense emotion. It occurred to Richer to resolve the problem by consulting the respiratory tracings of the subject under experiment. He effected the contraction of the muscles which express terror, and, strange to say, while the features and gestures of the subject expressed the most lively alarm, the breathing, after one abrupt act of expiration, resumed the calmness and immobility characteristic of catalepsy. Hence we may infer that in the cataleptic subject suggestions by the muscular sense are more superficial than the suggestions of somnambulism.

The chief conclusion to be drawn from these studies is the influence exerted on psychical activity by the expressive movements of the countenance and of the whole body. The expression is not merely an external sign of the emotion, but it forms an integral part of it. Even in the normal state, when an expression is artificially produced, it gives rise to the corresponding emotion, which passes away when the expression changes. Dugald Stewart's remarks on this subject have been often quoted. He observes that just as every mental emotion produces a sensible effect on the body, so when the countenance assumes the expression of any strong emotion, accompanied by analogous gestures, the emotion corresponding with this artificial expression is in some degree felt. Burke asserted that he had often experienced the awakening of the passion of anger in proportion as he assumed the external signs of that passion, and Stewart did not doubt that in the case of many individuals the same experiment would afford the same results. Burke also remarked that when Campanella, a celebrated philosopher and great physiognomist, wished to know what was passing in the mind of another person, he imitated, as well as he could, the attitude and countenance of the person in question, and at the same time he concentrated his attention on his own emotions.

Suggestions of attitude constitute the simplest form of automatism. A given number of co-ordinated movements may moreover be produced in some cataleptic subjects by placing their limbs in a certain position. For instance, if his hand is approached to his nose, the subject will blow his nose. Or certain impulses, which may be indefinitely prolonged, may be impressed on his limbs, as when the subject is made to twirl his thumbs, he will continue this automatic movement until it is stopped mechanically or by suggestion. These phenomena are simply automatic, and can be explained in great measure by the laws of the association of movements.

A higher form of automatism consists in what Heidenhain calls imitative automatism. The experimenter begins by looking fixedly at the subject, so as to arrest his gaze, and then draws back. The subject rises to follow the experimenter, of whom he never loses sight, and he imitates his every movement, whatever it may be. In this way he can be made to laugh, whistle, sing, blow his nose, clap his hands and feet together. The subject reflects the acts of the experimenter as a mirror might do; he imitates with his right hand the movements of the experimenter's left hand, who stands opposite to him. Despine has termed this phenomenon *specular* imitation.

Automatism may also be produced by the recollection of the use of an object. This is a less direct process, and the automatism is more complex. For instance, if a cake of soap is placed in the hands of a cataleptic subject, he rubs his hands, as if in the act of washing them. If an umbrella is given to him, he opens it, shivering, as if sensible of the coming storm. Sight, or contact with the object, automatically arouses a series of movements which are in the normal state associated with the same sensory impression. The subject relies on a basis of habits, and invents nothing. An unknown object produces no suggestion.

This is not the place for insisting on the analogies which exist between the movements, acts, and suggested ideas of hypnotic subjects and the spasmodic movements and ideas of the insane. We will only remark that in the case of the latter, an impulsive act is not unfrequently induced by the sight of an appropriate object. Max Simon gives an instance of a learned man who was seized by an overwhelming impulse to cut his throat when he was shaving, and who could only overcome it by desisting from this operation. Many other analogous facts might be cited.

Certain acts which are not purely mechanical cannot be suggested merely by the presence of the instrument which effects them. The act of writing, for instance, not only involves the exercise of the hand which traces the characters, but of the thought which co-ordinates the words in a given sequence. If, during the cataleptic state, a pen is placed between B——'s fingers, she holds it loosely, and at the end of a few moments lets it drop, without making any attempt to use it. If a sentence is dictated to her while she holds the pen, a word at a time, or still better, a syllable at a time, she may be induced to write a phrase or two in her own orthography, but the writing is irregular, since it is due to the influence of successive suggestions which are disconnected in the subject's mind. Yet if care is taken about the position of the hand, an autograph may be obtained which can hardly be distinguished from those which are written in the waking state.

There are several proofs of the absence of design,

The subject to whom soap is given will go on washing his hands indefinitely, and on one occasion the operation was protracted for two hours (Regnard). If a subject is putting on her boot, she will go on doing up and undoing the laces for an indefinite period, and if a piece of crochet work is given her, she will make a long chain of loop-stitch without attaching it to the rest of the work. Sometimes the act which is begun is continued indefinitely, owing to contact with the object which suggests the idea of employing it. More frequently, when the suggestion is exhausted, the subject stops short and becomes rigid in catalepsy. A species of oscillation may be observed between the cataleptic attitude and the psychical phenomena produced by suggestion. While the subject is affected by the suggestion, catalepsy ceases; as soon as the suggestion comes to an end, catalepsy reappears.

During the automatic activity it is possible to affect the subject unilaterally. Take a subject, for instance, before whom are placed a jug and basin and some soap. As soon as her eyes are attracted to these objects the subject, with apparent spontaneity, pours water into the basin, takes the soap and washes her hands with scrupulous care. If one of her eyes is then closed, the same side of the body becomes lethargic, while the other hand continues to exert the same movements. So, again, when the subject is working crochet and one eye is closed, the corresponding hand becomes motionless, while the other continues to exercise the same movements alone, although they are rendered useless by its isolation. Yet, as Richer observes, the intelligence seems to take some part in the unilateral movements; the subject tries to supply the place of the missing hand by supporting the other on the knee or breast.

None of these phenomena are peculiar to catalepsy: they may all be readily reproduced in somnambulism. But in catalepsy the movements are simpler and more automatic; the impulse seems to be irresistible.

It may be profitable to consider the facts of automatic imitation which are so readily produced in the cataleptic state; the echoing of any utterance, or *echolalia*. It is long since pathologists became acquainted with this phenomenon, which was discovered by Berger in the case of hypnotic subjects. It is produced in a somnambulist by applying the hand to the forehead or to the nape of the neck; the subject, who up to that time has answered the questions put to him with distinctness, at once repeats, instead of replying to the questions, as if he were transformed into a phonograph. He may be made to sing, scream, cough or sneeze; he will repeat words uttered in languages unknown to him, with an exactness which is often surprising. Some subjects also retain tunes, and may be made to sing a musical air; if a vibrating tuning-fork is applied to the ear, the subject reproduces the sound, with its pitch and quality. In this state also the subject automatically imitates all the gestures of the experimenter (Charcot).

Marie and Azoulay * have measured the period of reaction in echolalia. They adopted the following arrangement. A telephone was applied to the subject's ear, and his mouth was provided with a mouthpiece, so constructed that when the word "toc" was uttered by the subject, an electric signal was given by Marey's

* *Soc. de Psychologie*, May 18, 1885.

tambour. At the other end the fixed telephone was inserted in a circuit which included an electric contact and one of Deprez's signals, also registering on the same cylinder. Thus, when the electric contact took place, a sound in the telephone and a signal on the cylinder were simultaneously produced. The subject said "toc," whenever he heard the sound in the telephone, so as to give the period of personal reaction, as far as his auditory impressions were concerned.

In the waking state this period was $\frac{39}{105}$ of a second.
In the somnambulist state............ $\frac{33}{100}$ „
In echolalia .... $\frac{31}{100}$ „

II.

Acts only differ from movements in their complexity. Acts consist of associated movements, adapted by the subject to the end which he has in view. We wish to study these acts in the phase of somnambulism.

Verbal suggestion is the process usually employed. Heidenhain observed that when he said to his hypnotized brother, "If I had a watch, I should like to see what o'clock it is," no effects ensued. But if he said, "Show me your watch," the order was at once obeyed. We have been successful in giving such orders in writing. As soon as the subject read the words: "I am going to rise," he arose. In short, the only necessary condition is that the image of the act in question should be distinctly formed in the subject's mind.

To give an idea of the mathematical precision with which the suggested act is executed on awaking, one of the present writers performed the following experiment.

We showed to the somnambulist an imaginary spot on a smooth surface, which we could only afterwards ascertain by means of careful measurement, and we ordered her to stick a penknife into this spot when she awoke. She executed the order without hesitation and with absolute correctness: a criminal act would have been as punctually executed.*

It is interesting to ascertain whether the subject who is actuated by an irresistible impulse behaves like an automaton subsisting on a basis of the past, on his memory and habits, or if, on the contrary, the subject is capable of reflection and of reasoning like a normal individual. This latter is more frequently the case. When care is taken to suggest a somewhat complex act, for the performance of which some combination is necessary, we may observe that the subject invents such combined expedients although they had not been suggested to him, and this inventive process shows that everything is not explained by comparing him to an automaton. For instance, it was suggested to a subject that she should poison X—— with a glass of ·pure water which was said to contain poison. The suggestion did not indicate in what way the crime was to be committed. The subject offered the glass to X——, and invited him to drink by saying, " Is it not a hot day ? " (It was in summer.) We ordered another subject to steal a pocket-handkerchief from one of the persons present. The subject was hardly awake when she feigned dizziness, and staggering towards X——, she fell against him and hastily snatched his handkerchief. When a similar theft was suggested to a third subject, she approached X——,

* Ch. Féré, *Les hypnotiques hystériques*, etc.

and abruptly asked him what he had on his hand. While X——, somewhat startled, looked at his hand, his handkerchief disappeared. None of these expedients were suggested, but were derived by the subjects from their own resources. These complex phenomena cannot be referred to the simple fact that the image of a movement produces that movement; such a rudimentary explanation can only apply to elementary experiments.

There are numerous instances of resistance in hypnotized subjects. The order is disregarded, and not executed by the subject. The failure may arise from two different causes, derived either from the experimenter or from his subject. In the former case, as one of the present writers has already observed, the promptness and energy with which the act is performed depends on the authority with which the suggestion is given. When the order is given gently and indecisively, the subject awakes in a state of mind which it is interesting to study. She is uneasy, beset by the fixed idea that she has to do something which is absurd or revolting—to embrace a skull, for example. She hesitates long, and sometimes even expresses her hesitation, saying, "I must be mad, to wish to embrace a skull. It is absurd; I do not wish to do it, and yet it seems impossible to resist." And eventually she does it. It should be added that the personality of the experimenter has some share in the efficacy of the suggestion; a subject may resist an order given by one person, and obey the same order given by another. These facts were known to the early magnetizers, and they recommended the experimenter to use an authoritative tone, and the subject to be perfectly submissive.

At another time the subject's resistance may be due

to the nature of the suggested act. This resistance may be said to be a survival of the subject's personality; his personal reaction which is not completely destroyed by the hypnotic sleep. Such resistance occurs most frequently in those affected by profound hypnotism, and it is more common in some states than in others. We have already observed the automatism of the somnambulant state is much less absolute than that of catalepsy; the cataleptic subject is a machine, the somnambulist is a person. The first readily performs all the acts suggested, while the second often offers a resistance which may become troublesome to the experimenter.

Many subjects display their honesty by refusing to commit the thefts suggested to them. They assign various motives for the refusal. Sometimes the subject may reply: "No, I will not steal; I am no thief." Sometimes the motive is not so high. Many subjects reply to the suggestion by saying frankly, "Some one will see me." The suggestions of murder may provoke similar objections. If Z—— is armed with a paper-knife, and ordered to kill X——, she says, "Why should I do it? He has done me no harm." But if the experimenter insists, this slight scruple may be overcome, and she soon says, "If it must be done, I will do it." On awaking, she regards X—— with a perfidious smile, looks about her, and suddenly strikes him with the supposed dagger. But neither this subject nor any other could be impelled to murder some unspecified person. Another of our subjects presented an interesting example of invincible resistance. She had been deeply attached to a young man, and although she had suffered much from him, the passion was not extinct. If the presence of this man was evoked, she instantly displayed signs of great distress, and attempted to escape; but it was impossible to induce her to do anything which might be injurious to him whose victim she had been. Yet she automatically obeyed every other command. Another subject cannot be induced to repeat a prayer; a second will not sing a song which she has composed, reflecting on one of the present writers; a third resists the order to sign a cheque for a million francs, and will only do so for a much smaller sum.

Some hypnotized persons have the illusion of resistance, and believe that they can resist if they please. These illusions do not occur when the sleep is profound, and we have not met with them in our subjects, but Richet has observed them in some cases. He writes: "One of my friends, who was drowsy but not quite asleep, carefully studied this phenomenon of incapacity, combined with the illusion of capacity. When I prescribed a movement, he always performed it, even although he had, before he was magnetized, been determined to resist. He found this hard to understand when he awoke, and said that he certainly could have resisted, only he did not wish to do so. Sometimes he was inclined to believe that he was simulating. 'When I am asleep,' he said, 'I feign automatism, although I believe that I might act otherwise. I begin with the firm determination not to simulate, but as soon as I am asleep it seems that, in spite of myself, simulation begins.' It is evident that this mode of simulating a phenomenon does not differ from the real phenomenon. The automatism is proved by the simple fact that in all good faith persons act like automata. It matters little that they believe themselves capable of resistance, since as a fact they do not resist. This is what we have to consider, not their illusion as to their imaginary power of resistance." *

Among the psychical phenomena which accompany the motor impulse that is suggested, there is perhaps none more interesting than the apparent motives which the subject assigns to his act. These facts show, as the illusion of resistance has already shown, that the subject is altogether ignorant of the original source of the impulse he has received. When the subject awakes, and performs the act which was suggested to him during somnambulism, he generally supposes it to be spontaneous; the suggested act, imposed by the will of another, seems to him of precisely the same nature as those acts which he performs of his own initiative. And again, since he is unaware of the true cause of action, the subject invents a motive, more or less plausible or ingenious, to explain to himself the reason of his conduct. Richet was the first to make a regular study of this phenomenon, and we give some of his observations: "When B—— was hypnotized, I said to her: 'On awakening, you will take the shade off the lamp.' I awoke her, and when we had conversed for a few minutes, she said: 'We do not see well,' and she took off the shade. Another time I said to B——, 'When you awake, you will put a good deal of sugar in your tea.' I awoke her, tea was served, and she filled her cup with sugar. Some one asked what she was about. 'I am putting in the sugar.' 'But you put in too much.'—'Really! that is a pity.' And she continued to put it in. Then she said, on finding

* Richet, *L'homme et l'intelligence.*

the tea undrinkable: 'What would you have? It was a stupid thing to do; but have you never done anything stupid?'" *

An analogous observation was made by one of the present writers. Together with M. B——, whose first visit to the Salpêtrière was made on that day, we performed some hypnotic experiments on one of Charcot's hystero-epileptic patients. When the subject was in a state of somnambulism, I ordered her on awakening to stab M. B—— with the pasteboard knife I put into her hand. As soon as she awoke, she rushed towards him and struck him in the region of the heart. M. B—— feigned to fall down. I then asked the subject why she had killed this man. She looked at him fixedly for a moment, and then replied with an expression of ferocity, "He is an old villain, and wished to insult me." †

It is evident that there was in this case no substantial motive for the crime, nor had any motive been suggested to her. When the crime was accomplished, the subject hesitated for a moment before assigning a motive; her conscience was at fault, and she questioned the outward aspect of her victim. No great power of observation was necessary in order to note M. B——'s sprightly expression of countenance, and this was enough to supply the answer. He had not struck nor robbed her, and since there was no other reason for stabbing him, an insult must have been the cause, since she would not have done such a thing without a reason. It must

* Richet, *La Mémoire et la personnalité dans le somnambulisme* (*Revue Philosophique*, March, 1882).

† Ibid.

be remembered that epileptic subjects, when they have involuntarily committed a crime may, like the subjects of suggestion, not only admit their guilt, but explain it by more or less rational motives. This is only another proof that experiments in hypnotism are valuable in the treatment of mental diseases.* We may add that a suggested impulse resembles the irresistible impulses of some insane persons in two important features: the subject's anguish when he is restrained from accomplishing the act, and his relief when it is accomplished.

Suggestions which are not to be at once acted upon are possible in the case of acts as well as of hallucinations. Richet was the first to call attention to these experiments.† "When B—— was hypnotized," he writes, "I said to her: 'You will come here on a given day and at a given hour.' When she awoke she had forgotten these words, and she asked when I wished to see her again. I said: 'Whenever you can come; any day next week.' —'At what o'clock?'—'Whenever you please.' And she came regularly, with surprising punctuality, at the date and time indicated by the suggestion. This phenomenon sometimes leads to absurd consequences. A—— arrived one day at the hour agreed upon during hypnotism, and the first thing she said was: 'I do not know why I came. The weather is horrible, and I had visitors. I had to run to get here in time, and I cannot stay. I must go back in a few moments. It is absurd, since I do not understand why I came. Is this another phenomenon of magnetism?'"

* Féré, *Note pour servir à l'histoire des actes impulsifs des épileptiques* (*Revue de Médicine*, 1885).

† *Revue Philosophique*, March, 1888.

These experiments give rise to the same difficulties as the hallucinations which occur at a fixed date, and we need not go over the question again. The only point peculiar to suggestions of command is that up to the moment which has been fixed, the subject does not perform the act enjoined, even when led up to it and reminded of the order. Suppose that a subject is told that at five o'clock on the following day he will read page 8 of this book. On awaking, the book is presentĕd to him, open at the page just indicated, but it suggests nothing to him. The suggestion is only realized at the given hour, and cannot be realized until that hour arrives.*

It cannot be denied that all these facts have disquieting consequences with respect to the existence of free-will. Psychologists of the spiritualist school have long regarded the sense of liberty with which we all perform a voluntary act as a proof of free-will. The history of suggested impulses show what is the value of this subjective sense, which has been exalted into an objective proof, and which is perhaps only an illusion. Philosophers will have to ask themselves what confidence can be placed in what Leibnitz termed "the lively internal sense of freedom," since this sense may be so greatly deceived. Spinoza's profound remark on this subject must be remembered: "The consciousness of free-will is only ignorance of the causes of our acts." It must be admitted that these words are perfectly applicable to the acts produced by hypnotic suggestion; the subject believes himself to act freely, because he has forgotten the suggestion by which he is impelled. It may be asked whether we can reason from a hypnotic patient to a man

* Beaunis, *Le Somnambulisme provoqué*, p. 57. Baillière, 1886.

of sane and normal mind. Some philosophers may accept this as a mode of escape, and we cannot go more deeply into the question. But we think that at any rate the experience of hypnotism proves one important fact, that the testimony of our inner consciousness is not infallible.

We have not yet dwelt upon the form in which the suggestion is given. For the most part the act to be performed is simply indicated : "When you awake you will clap your hands." The wish to perform the act may also be suggested. "You are very angry with X——, and when you awake, you intend to strike him." A suggestion of incapacity may also be substituted for the suggestion of will: "I order you to strike X——, and however much you resist, you will be obliged to obey." In all these cases the result is the same, and the act suggested is performed. From the psychical point of view there is, however, a wide difference between the agent who performs an action because he wishes it, and the agent who obeys the irresistible will of another person. Yet hypnotism shows that this difference is merely superficial. In both cases there is what may be called in psychological terms the same impulse, and in physiological terms the same dynamic state of the motor centres. We were confronted with a similar fact in the study of hallucinations. Hallucination, memory, and sensation, as we then observed, are clearly founded on the same physiological act, which takes place in the same region of the cerebrum. They are only distinguished by the secondary states of consciousness which accompany the formation of the image. In memory, this state consists in the reasoning which localizes the image in the past. In hallucination and in sensation, these states consist in the reasoning which localizes the image in the external world. But these localizations in time and space are superadded acts, which are not essential and are often absent. We believe that it is the same with volition. The impulse is the fundamental fact, around which may be grouped the secondary states of consciousness which make the impulse a voluntary or involuntary act, or which assign to it a given motive. These are accessory and superadded phenomena, not integral parts of the occurrence.

Finally, we must indicate the relation which exists between cataleptic attitudes and the attitudes produced by a suggestion given during somnambulism. By suggestion a subject may be induced to maintain a given attitude for some time, as he does during catalepsy. This retention of attitudes under the influence of an idea cannot last indefinitely; its duration depends upon many circumstances, and chiefly upon the muscular strength of the subject, and on the form of the suggestion. If the subject is merely ordered to keep his arm horizontally extended, the arm soon begins to tremble, and respiration becomes irregular. In one debilitated subject the trembling was very marked, and the arm dropped in two minutes. But if this subject was told that her arm was made of wood, then the extended arm did not precisely tremble, but was affected by slow oscillations which moved the whole limb, and it only dropped at the end of three minutes. Consequently the attitudes imposed on our subjects by suggestion differ from those impressed upon them during catalepsy, a difference which proves that the catalepsy of these subjects is not a state produced by suggestion. But we do not wish to assert that by subjecting patients to repeated experiments in suggestion, it might not be possible to give them attitudes resembling those of true catalepsy.

III.

We have submitted to the action of the magnet the unilateral form of the acts and movements suggested during hypnotism, and we have ascertained that these unilateral phenomena may be transferred like hallucinations, and other physical symptoms of hypnotism.[*]

After hypnotizing one of our subjects, we placed a bust of Gall on a table standing near. We then suggested that she should *make a long nose* at the bust with her left hand. A magnet was placed close to her right hand. On awaking, the subject, as soon as she saw the bust, made a long nose at it with her left hand; after the lapse of a few seconds she began again, and repeated the gesture fourteen times, always with the left hand. The latter movements were more faintly executed, and the gesture was not fully carried out; she only raised her hand as high as her mouth, without extending the fingers. A slight tremulous motion then began in the right hand, and the left hand remained still. The subject appeared to be uneasy, and turned her head from one side to another; she addressed the bust, saying: "How offensive that man is!" She scratched her ear with the *right* hand and then began a series of the same gestures as before with that hand,

* *Revue Philosophique*, January and March, 1885.

which went on for ten minutes. She admitted that it was absurd to make such gestures, yet if she paused for an instant, it was only necessary for the experimenter to make a long nose at the bust to cause her to resume them. We withdrew the magnet, and the action was then transferred to the left hand, with the same characteristics as before. We gave the subject a piece of work to employ her hands, but she laid it down at regular intervals of three or four seconds, in order to make a long nose. From time to time she complained of pain, which oscillated from one side of the head to the other.

This is an instance of a transferred act, which was suggested during somnambulism, and yet had all the appearance of being spontaneous. The subject invented specious reasons to explain her conduct; she said that the bust was offensive, and believed that she made a long nose at it for this reason. As we have observed, when she paused, it was only necessary to imitate the gesture in order to *re-charge* the subject, and make her resume the insult. This proves the force of example, or rather, the influence of the representation of the movement on the movement.

The ensuing experiment defines the result of the transference, and shows that the transferred act is symmetrical with the suggested act. We impressed on a hypnotized subject the idea of setting down figures in the ordinary way, with her right hand, and at the same time a magnet was concealed near her left hand. On awaking, she wrote as far as the figure 12 with her right hand, then she hesitated, changed the pen to her left hand, and went on writing. The figures she set down were correct when seen in a mirror, so that the movements executed with the left hand were symmetrical with those made with the right hand. The magnet had therefore transferred the action. It should be observed that while she was writing with the left hand, she was unable to use the right; she was as it were left-handed with her right hand.

Fig. 13.—Experiment on June 16, 1884.   Magnetic transfer of impulse to write.

Fig. 13 represents the first experiment of the transfer of writing. These figures were set down with the left hand. Only the three first figures are reversed.

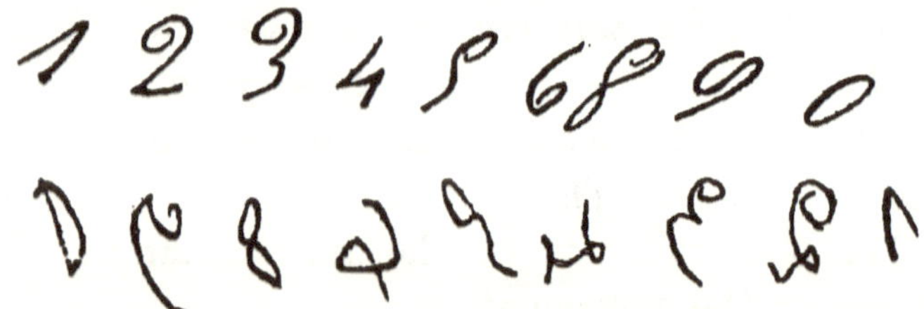

Fig. 14.—Experiment of November 29, 1884.   Magnetic transfer of impulse of writing.

Fig. 14 represents a subsequent experiment. The subject had improved; the first line of figures was written with the right hand, and that below with the left, set down from left to right. The figure 7 is absent, because it had been suppressed by suggestion in a previous experiment. The reversed writing produced by the magnet demands attention. This phenomenon is due to the fact that the magnet has transferred the impulse to set down figures from right to left and reversed writing is the normal writing of the left hand. This fact has been proved by many experiments.

The transfer of verbal impulse, which is in fact only a variety of the motor impulse, may also be effected. We suggested to a subject that she should count aloud, up to 100. She began to count as soon as she awoke. A ten-forked magnet was placed near her right arm. When she got to 72 she paused, hesitated, could not go on counting, and at the end of a minute she was unable to speak at all. Yet she could move her tongue, and understand what was said to her. After ten minutes had elapsed, the magnet was applied to her left side; in about two minutes the left arm began to tremble and the power of speech returned.

Finally, let us note the transfer of resolutions, that is, of proposed, but unfulfilled acts. We said to X—— when under somnambulism: "Here is the key of the wardrobe at the end of the room. When we offer you the key, you will take it in your right hand; you will open the drawer, take out a box, close the drawer, and finally give the box to B——: all with the right hand." On awaking the subject, a magnet was placed near her right fore-arm. After a few moments X—— complained of pain in the right parietal region; pain which traversed the head and passed into the same region on the left. A minute afterwards we offered her the key; she took it in her left hand, walked to the wardrobe and attempted to open it with her right hand, but was unable to do so. She had recourse to her left hand in order to open the drawer, and she went through the same process before taking the box; she alternately extended the right and left hand, and finally made use of the latter. She closed the drawer, after the same hesitation with the left hand, came back with the box, stood before B——, and said, " Here is the box, sir," and offered it to him with her left hand.

We repeated this experiment a second time, allowing five minutes to intervene before presenting the key, so that the transfer might be completely effected. In this case the subject took the key in her left hand without the slightest hesitation, she opened the drawer, took out the box, closed the drawer and gave the box to B—— without any attempt to use the right hand.

We have here a peculiar kind of transfer. There is a resolution to perform an act, which is in some sense present in the subject's cerebral cells. This virtual act is susceptible of transference, precisely like an act which is actually accomplished, and this clearly shows that it has likewise a material substratum. We should also observe the phenomenon of pain which accompanies the transfer; a pain which is not diffused but local, and according to the theories of cranio-cerebral topography already established by one of the present writers, the pain is localized in the ascending frontal and parietal convolutions, in the motor centres of the limbs. It will be remembered that the pain which accompanies the transfer of an act has the same localization. This resemblance seems to show that the resolution to perform an act with a given limb, with the right arm for example, corresponds with a physiological process which has the same site as the movement of the arm. The potential act—and a resolution to act is nothing else—seems to have the same cerebral centre as the act which is really performed.   It must be understood that we give this interpretation with all reserve, and that it is merely an hypothesis.

Finally, the magnet exerts its special action on spontaneous phenomena, which have not merely the appearance of being freely displayed, but are really voluntary, in the ordinary sense of the word.   In fact transference may be effected without the hypnotic sleep or suggestion. The subject is merely requested to perform a given act, and the application of the magnet will compel him to perform a second act, symmetrical with the former.   We give the experiment without further comment.

X—— was perfectly awake, and had not been hypnotized for several days. We begged her to rest her right elbow on the table, close to a concealed magnet. She asked the reason, and we made the pretext of a wish to take her portrait, to which she agreed. After two or three minutes, she brought her right arm close to her body, saying that she was tired and that her arm was numb. She seemed uncertain for a moment, and looked to the right and left. We begged her to resume her position, and she said she had forgotten what it was; in another minute she rested her left elbow on a chair she had drawn up, in a position symmetrical to the former one.   When the magnet was withdrawn, consecutive oscillations were observed.

The magnet will likewise effect a bilateral act: the result differs according to the simply automatic character of this act, or its correspondence with the emotional state; in the former case it produces what we have

14

called a motor polarization, and in the second case an emotional polarization.

We first give an example of motor polarization. The idea is suggested to a subject that he should move both his hands as if rolling a ball. While he is continuing this action, a magnet is brought close to the nape of the neck. After a while both hands begin to tremble; the subject tries to roll his fingers without being able to do so, and seems at a loss what to do. The suggested impulse is succeeded by a corresponding paralysis. Many other experiments might be cited, in which the impulse is likewise changed into paralysis.

When motor polarization is compared with sensory polarization, it will be seen that they are at once alike and unlike. When the vision of the colour red is polarized, three effects are produced: suppression of this vision, paralysis, as far as red is concerned, and the subjective consciousness of the complementary colour, green. When an automatic movement is polarized the two effects of suppression and of paralysis are indeed produced, but the production of an inverse, complementary phenomenon corresponding with the colour green, appears to be absent. This gap is more apparent than real, and we shall presently attempt to fill it.

We must now give an instance of emotional polarization. We impressed the idea on a hypnotized subject that she would on awaking feel a desire to strike F——. A magnet was placed near her right foot. As soon as she awoke, she looked uneasily at F——, got up suddenly, and tried to give him a slap which he had just time to ward off. "I do not know why," she said passionately, "but I feel a desire to strike him," and indeed she tried hard to do so. In another moment her countenance changed, she assumed a gentle and supplicating expression, threw herself on the experimenter, saying, " I want to embrace him," and it was necessary to use force to restrain her impulse. Consecutive oscillations were then observed.

In this last experiment, the magnet directly polarized the suggested emotion, which in its transformation led to a fresh series of acts. This is an emotional, not a motor polarization, a distinction which should be clearly understood. The magnet, if it acted solely on a motor phenomenon, such as the act of striking, would not substitute for it phenomena of another order, such as the act of embracing: for the opposition of these two acts is due to the difference in the emotion they express, and not to the difference in their motor character. The state of emotion is therefore the pivot on which the experiment turns.

The analysis of this emotional polarization will show that it consists of the three elements mentioned above: suppression, paralysis, and manifestation of the converse state. When we compare these facts with those obtained from the polarization of colours, we shall see that there are complementary emotions, just as there are complementary colours.*

---

* For further details, see *Revue Philosophique*, March, 1885. Bianchi and Sommer (*Archivio di psichiatria, scienze penali,* etc., vol. vii. p. 387; 1886) were successful in reproducing some of the phenomena of psychical polarization which were discovered by the present writers.

CHAPTER XI.

PARALYSIS BY SUGGESTION : ANÆSTHESIA.

THE study of the different forms of paralysis by suggestion opens a perfectly new horizon to psychology, which
is not bound by the mental laws hitherto established,
and refuses to be included within the too restricted limits
of its classification. If you consult one of the classic
works on psychology, you will find in it the three great
divisions of emotion, intelligence, and will, and in none
of these does psychical paralysis find a place. In fact
the very name of psychical paralysis is new. Up to this
time it has only been used metaphorically, and the notion
of this fact has only been slightly indicated by writers
on the subject. The experimental method employed in
hypnotism was necessary to reveal the existence and
extent of paralysis by suggestion. We are now aware
that it may affect all the parts of the psychical mechanism,
sensation, imagination, memory, reason, will, motor power,
etc.; it is, in a word, co-extensive with the intelligence.
Classical psychology, which does not mention psychical
paralysis, omits half the history of the mind: it describes
the active, impulsive forms of the intelligence, without
taking note of the passive, negative forms, which are
equally numerous; it represents that side of the mind

on which the light falls, without taking note of the side in shadow.

The course of this work has repeatedly brought us in contact with paralysis by suggestion. We have seen that when an æsthesiogen acts upon a bilateral hallucination, the latter is destroyed, and that it is succeeded by a corresponding anæsthesia. So, again, when a bilateral movement is subjected to magnetic action, the corresponding paralysis is produced. It may also be observed that when suggestion puts an end to an hallucination or an act, these active phenomena give place on disappearing to a paralysis representing their negative form. We have now to consider these forms of paralysis, as they are displayed in consequence of a direct suggestion.

Sensibility can be destroyed by suggestion. This fact of anæsthesia by suggestion has long been known, and has sometimes been employed in cases of surgical operations. One of the present writers was able, by means of suggestion, to open an abcess, seated in the axilla, without causing pain. There is nothing more surprising than this power of destroying pain by suggestion. The anæsthesia may be such as to lead the subject to believe that the limb is gone; it may affect, not merely the general sensibility of the body but the special senses. It would be easy to render some subjects perfectly blind by suggestion, but the operator must prudently refrain from such serious experiments, lest he should be unable to put an end to their results.

We propose to make a special study of systematic anæsthesia, erroneously termed by Bernheim and some other writers negative hallucinations.

Since the definition of systematic anæsthesia presents

pecial difficulties, it seems well to defer it, and provisionally to substitute for it as complete a description as possible. We suggested to a hypnotized subject that when she awoke she would be unable to see F——, but that she would continue to hear his voice. When she awoke, F—— placed himself before her, but she did not look at him, and when he extended his hand, there was no corresponding gesture on her side. She remained quietly seated in the chair in which she had been sleeping, and we sat waiting beside her. After a while, the subject expressed surprise at no longer seeing F——, who had been in the laboratory, and she asked what had become of him. We replied, "He has gone out; you may return to your room." F—— placed himself before the door. The subject arose, said good morning, and went towards it. Just as she was about to lay hold of the handle, she knocked up against F——, whom she was unable to see. This unexpected shock made her start; she tried to go on again, but on encountering the same invisible and inexplicable resistance, she began to be afraid, and refused to go near the door.

We next took up a hat, and showed it to the subject. She saw it quite well, and touched it in order to satisfy herself that it was really there. We then placed it on F——'s head, and words cannot express the subject's surprise, since it appeared to her that the hat was suspended in the air. Her surprise was at its height when F—— took off the hat and saluted her with it several times; she saw the hat, without any support, describing curves in the air. She declared that it was *de la physique,* and supposed that the hat was suspended by a string; she even got upon a chair to try and touch

this string, which she was unable to find. We then took
a cloak and handed it to F——, who put it on. The
subject looked at it fixedly with a bewildered air, since
she saw it moving about and assuming the form of a
person. "It is," she said, "like a hollow puppet." At
our command the furniture was moved about and noisily
rolled from one end of the room to the other—they were,
in fact, displaced by the invisible F——; the tables and
chairs were overturned, and then the chaos was succeeded
by order. The different objects were replaced, the dis-
jointed bones of a skull, which had been scattered on the
floor were joined together again; a purse opened of itself,
and gold and silver coins fell from it.

We then induced the subject to sit down again, and
we placed ourselves beside her chair, in order to subject
her to experiments of a quieter nature. We shall see
how she managed to explain certain facts, rendered
inexplicable by her inability to see F——. That gentleman placed himself behind her, and while she was
quietly conversing with us, he touched her nose, cheeks,
forehead, or chin. Each time the subject put her hand
to her face in a natural way, and without any appearance of alarm. We asked why she put her hand to
her face, and she replied that it itched, or was painful,
and she therefore scratched it. Her tranquil assurance
was extremely curious. We begged her to strike out
violently into space, and at the moment she raised her
arm it was arrested by F——. We asked what was the
matter, and she replied that her arm was affected by
cramp. She was, therefore, never at a loss; she invariably
explained everything, however insufficient the explanation might be. This need of explanation, which exists

in the normal state, is carried to excess in the experiments produced by suggestion.

Such are the main lines of the phenomenon of systematic anæsthesia, and it may be well to insist on some points of this description. With respect to the extent of the systematic anæsthesia, it should be said that when a small object, such as a pencil, is rendered invisible, it is this object alone which the subject is unable to see. The limits of anæsthesia are fixed, its extent invariable. This is not the case when the perception of a more complex object is destroyed; the anæsthesia then affects all which is indirectly connected with the object. If suggestion has rendered a purse invisible, the subject may fail to see the coins which issue from it. When it is a person who is rendered invisible, the subject cannot see the person, nor the clothes he wears, nor—which is more curious—the things he takes out of his pocket, a handkerchief, watch, or key. But these results are very variable, and differ with different subjects. Although the perception of the object is destroyed, it cannot be said to be as though it were non-existent, for its presence continues to be displayed by certain signs. For instance, when it has been suggested to the subject that he cannot see the light, the pupils continue to react when his eyes are turned to the window. So, again, we have seen that when suggestion has rendered the magnet invisible, transference and polarization can be effected in some subjects. And again, a person rendered invisible by suggestion may hypnotize a subject by means of passes. Moreover, the suppressed object may continue to act on the conscious sensibility of the subject. Suppose that a scent-bottle is rendered invisible, and the subject is told: "You will no longer see this bottle; your fingers will not feel it; when you strike it, it will not sound; it will have no existence for you." The subject receives the scent-bottle on awaking, and it is true that he does not see it, nor feel its contact. But an intelligent subject will soon perceive that there is something in his hands, and be conscious of resistance when he tries to put them together. One of our subjects, after studying the nature of this resistance, came to the conclusion that the object was round, and this offers a curious analysis of the sense of touch, and of the muscular sense.

In order to complete the description, we should ascertain whether the object which has become invisible conceals what is placed behind it. This is sometimes the case. If F—— puts on a pince-nez when he is invisible, and then turns his back on the subject, she can no longer see the pince-nez. But generally the invisible object does not prevent the subject from seeing the things beyond it in the same line of vision. It does not cause an apparent gap in the field of vision. The subject believes that he sees the hidden object as well as the rest. When F—— stands before the door, the subject maintains that she still sees its handle, and will try to . take hold of it. It is probable that she spontaneously creates an hallucination in order to fill up the gap which the invisible object has produced in her field of vision. This effect of auto-suggestion recalls a well-known physiological fact—the normal existence in the field of vision of a gap or blind spot, corresponding with the entrance of the optic nerve. The existence of this spot is only ascertained by means of experiments, since it is filled up unconsciously during normal vision. It is scarcely necessary to add that, in spite of appearances to the contrary, the invisible object really acts as a screen, and that the subject cannot see what is on the other side of it. If we stand behind F—— when he is invisible, the subject maintains that she can still see us, but she cannot accurately describe our gestures.

We must conclude this superficial description of systematic anæsthesia by proving its reality, since all the preceding phenomena might result from adroit simulation; they do not include any of the material, objective characteristics which completely exclude the suspicion of fraud. But there is a mode of ascertaining the good faith of the subject. We know that the deafening noise of a Chinese gong produces catalepsy in some hypnotic subjects, and among these must be included two who were submitted to our observation. During the hypnotic sleep we impressed upon them the idea that they would no longer see the gong nor its sounder on awaking, and that they would be unable to hear its noise. In other words, we suggested a systematic anæsthesia, of which the gong was the object. When they awoke, and the maintenance of the anæsthesia had been ascertained, the gong was brought close to their ears, which they permitted without displaying the terror it habitually excited, and it was violently sounded. No catalepsy ensued, and in each case the subject did not flinch. She made a slight movement of surprise, and said that she had heard something like a gust of wind in the chimney. The counter-proof was immediately afforded by again hypnotizing the subject and restoring to her the perception of the instrument. A much fainter stroke then threw her into a profound catalepsy. In a fresh experiment we destroyed the existence of the gong, and for eleven successive days its effects were absolutely negative, after which the effect of the suggestion spontaneously disappeared.

Another experiment indicates that these forms of anæsthesia are genuine. One of our subjects had, on each side of the mammary regions, an hysterogenic zone, pressure on which immediately produced an hysterical attack. One of the present writers rendered himself invisible by suggestion, and at the same time destroyed the sensation of contact on his approach. A strong pressure of the hysterogenic zones then failed to produce any attack in the subject, nor did she make any effort to repel the experimenter: she only complained of a vague sense of oppression. On the other hand, she recoiled in terror when another person put his hand near these zones.

The researches we have made into the duration of systematic anæsthesia are still incomplete. It is often maintained for several days. In the case of our hysterical patients, the effect of the suggestion is destroyed by an attack. We have, however, repeatedly observed instances of anæsthesia with reference to small objects, such as watches, pencils, etc., which lasted for several months. Their mode of disappearance is remarkable. When a person has been rendered invisible, the subject does not see nor recognize him, but as time goes on the anæsthesia gradually becomes fainter. It is a curious fact that when the subject begins to see the person in question, she fails to recognize him, and the act of recognition only occurs later, by a species of ascendant evolution. Thus F——, who was the object of experiment in the observations

we have cited, was visible to the subject after the lapse
of three or four days, but she did not recognize him :
she took him for a stranger who had come to see the
Salpêtrière.   We have noticed elsewhere the physiological importance of these phenomena.*

We recently observed the converse of the preceding
fact in the case of a subject named C——, who was
subjected to an experiment in anæsthesia for the first
time. It was suggested to her during somnambulism
that when she awoke she would no longer see D——,
one of the persons present, whose name she knew. When
she awoke she saw, but did not recognize him, and at
the same time she had forgotten her own name and
identity. This subject had received, a few minutes before,
the hallucination of D——'s portrait on a blank card.
She was now made to look at the card, and after
repeatedly comparing the portrait with its original,
she became able to recognize D——. The systematic
anæsthesia was therefore destroyed by a recollection,
just as in some cases paralysis may be destroyed by
recalling the movement to mind.

It would be an error to suppose that systematic
anæsthesia consists solely in a sensory disturbance. In
the case of our subjects we have often found that this
suggested disturbance readily serves as the point of
departure for delirium. Thus, we once suggested to an
hypnotic subject that she would cease to see F——, but
would continue to hear his voice. On awaking the
subject heard the voice of an invisible person, and looked
about the room to discover the cause of this singular
phenomenon, asking us about it with some uneasiness.

* Binet, *La psychologie du raisonnement.* Paris, 1886.

We said jestingly, "F—— is dead, and it is his ghost which speaks to you." The subject is intelligent, and in her normal state she would probably have taken the jest at its true value; but she was dominated by the suggestion of anæsthesia, and readily accepted the explanation. When F—— spoke again he said that he had died the night before, and that his body had been taken to the post-mortem room. The subject clasped her hands with a sad expression, and asked when he was to be buried, as she wished to be present at the religious service. "Poor young man!" she said; "he was not a bad man." F——, wishing to see how far her credulity would go, uttered groans, and complained of the autopsy of his body which was going on. The scene then became tragic, for the emotion of the subject caused her to fall backwards in an incipient attack of hysteria, which we promptly arrested by ovarian compression. This experiment shows that when a subject remains under the influence of a suggestion after awaking, he has not, whatever be the appearances to the contrary, returned to his normal state. The suggestion of anæsthesia has disturbed the intelligence, and exerts a suspensory action on the judgment and on the critical sense.

In order to understand the nature of systematic anæsthesia, it is necessary to compare it with the spontaneous phenomena of hysteria, which resemble it. We frequently meet with paralysis of the senses in hysterical patients. There is one remarkable characteristic of hysterical anæsthesia, whether suggested or spontaneous, and this is especially apparent in the anæsthesia with respect to colour, or achromatopsia.

Suppose that the left eye of an hysterical subject

displays complete achromatopsia, extending to all colours. If a square of red is shown to her, while her right eye is closed, this square appears to her to be black or grey, and she has no conscious perception of red. Yet the red ray emitted by the object still exerts its special action on the visual centre of the subject, for if she is requested to close the right eye and to look fixedly for a moment at the red square which appears to her to be grey, she will after a while obtain the after-image of a green square. Thus the red, which the patient does not see, has enabled her to see the complementary colour, green, and, in spite of her achromatopsia, the colours produce correct after-images. This curious experiment, which was performed several years ago by Regnard, may be varied in many ways. For instance, it has been ascertained that, even when the subject is colour-blind, the invisible red, when mixed with a visible green, produces white, etc. From these facts Regnard drew the probable conclusion that hysterical achromatopsia does not, as the theory of Helmholtz asserts, result from a lesion of the elements of the retina, but from a modification of the centre of vision.

In order to ascertain whether the achromatopsia of suggestion displayed the same characteristics as true achromatopsia, we made a square of red paper invisible; we then requested the subject to look fixedly at its centre, and she saw the after-image of a green square. A repetition of the experiment with other colours always afforded corresponding results.* The mixture of un-

---

* We may observe in passing that in this anæsthesia by suggestion, the consecutive image retains the form of the actual image ; a red cross produces the subjective image of a green cross. On the other hand, in the

perceived colours produced the same tints as if their components were perceptible.   In short, we ascertained that, notwithstanding their diversity of origin, the charac-teristics of suggested and of spontaneous achromatopsia were the same.   The paralysis was of the same nature. It is, therefore, probable that when anæsthesia is suggested in the case of colours, the coloured ray reaches and pene-trates the sensory centre, since it produces a sensation of the complementary colour.   Suggested achromatopsia is a central, not a peripheral disturbance.

An experiment of another kind leads to the same conclusion.   A blank card is shown to the subject, and she is told that she will not see what is placed upon it. On awaking, the subject's attention is drawn to a square of blue paper, placed upon the card, which, however, still appears to her to be blank.   Yet it may be ascertained that although not consciously perceived by the subject, her brain, like a photographic plate, has registered all the modifications of the card, and this negative proof may subsequently be developed and rendered visible. In other words, the subject may receive a conscious recol-lection of the blue square laid upon the card.   This is most easily effected by the magnet.   The subject is invited to picture to herself the colour of the card, and at first it appears to her to be white, but under the influence of magnetization the centre of the card becomes darker, and she finally declares that she sees on it a small blue square.

The anæsthesia by suggestion which we are now con-

case of anæsthesia produced by the application of the magnet, the after-image is disfigured; a blank space in the form of a cross, and surrounded by green, succeeds to the vision of a red cross.   We are ignorant of the causes of these differences.

sidering displays one important feature—systematization. The spontaneous anæsthesia of hysteria is in some sense diffuse. When a patient is unable to see the colour red, she is unable to see that colour in the case of any object, whatever be its form and nature. On the other hand, a suggested anæsthesia may easily be produced which refers to a single, definite object. For instance, the subject may be unable to see one particular cross of red paper, while she still sees all the other figures, and even other crosses, cut out of the same paper. And again, suggestion prevents the subject from seeing X——, while all other people remain visible to her.

We wish to insist on the systematic character of suggested anæsthesia, since its consequences are singular, and we might even say paradoxical. Since suggestion only deprives the subject of the perception of a special object, it follows that the subject must recognize this object in order not to see it. This fact is illustrated by an experiment. We took one of ten cards which were apparently alike, and said to a somnambulist subject, " When you awake, you will no longer see this card." On awaking, we offered her the ten cards, one after the other, and she took each in turn except one, which she did not appear to see. It was the one to which we had directed her attention, and which we had rendered invisible. She therefore distinguished it from the others and recognized it, since she obeyed the suggestion which made it invisible.

It is clear that there is nothing supernatural in this recognition of an invisible object, when it is mixed with several similar objects. The subject was probably aided by some distinctive mark in which it differed from the others, and this was also the case when she picked out of ten cards the one on which suggestion had placed an imaginary portrait.   There is some analogy between these two experiments, since in both cases the optic image of the card, with all its slightest details, is photographed in the subject's brain, and serves as a guide to her researches.

It is curious that this recognition of the card, a complex and delicate operation, involving a sustained effort of attention, should end in a phenomenon of anæsthesia.   It seems probable that this act of recognition occurs altogether in the region of unconscious vision.   This explanation—admitting it to be an explanation—also applies to the fact that when a person who has been rendered invisible takes out his pocket-handkerchief, the subject does not see the handkerchief.   If he does not see it, it is clearly because he has ascertained that the handkerchief came from the pocket of the invisible person.   Here, again, we find an act of unconscious reasoning, which precedes, prepares, and directs the phenomenon of anæsthesia.

It sometimes happens that the attempt to produce systematic anæsthesia results in a shallower and less defined phenomenon, which is, however, interesting, since it approximates to the phenomena of normal life.

On awaking from the hypnotic sleep, six cards were placed on the table before X——, and she perceived and remembered their number.   One of the present writers took one of them, showed it to her, and asserted that it did not exist.   After some resistance, she finally admitted that his hand was empty.   But when she was told to pick up the cards and give them back to him in succession, it was not always the card which had been rendered invisible which she left upon the table; it was sometimes one card, sometimes another. Moreover, if a certain number of other cards was added to the six, without telling the subject how many they were, she counted them all from the first to the last, including the invisible card in her enumeration. This experiment seems to show that the suggestion made in the waking state had not produced in the subject a sensory anæsthesia with respect to a given card, but rather the fixed idea that there were only five cards upon the table. There was not the profound disturbance implied in an error of the senses, but it was an error of the reason, and it was this fixed idea which unconsciously inclined her to leave out one card when she attempted to count them.

There is something in normal life which closely resembles this attenuated form of anæsthesia. The preconceived idea that certain objects occupy a certain place is a hindrance to seeing them elsewhere, when they are displaced. We have often observed this peculiarity. If the hand of another happens to have removed an object which habitually stands on our study table, we look about for it, and may pass over the place in which it now stands ten times without perceiving it. Nor is this the only point of contact between the strange facts of invisibility by suggestion, and the known facts of normal life. It should be observed that these effects of psychical inhibition are produced in subjects by a negative form of suggestion. The experimenter always utters a negation, saying, " You do not see that person, you do not feel the contact of my hand." In other words, he impresses on his subjects the conviction that a given object does not exist. It may, therefore, be inferred that in the normal state, whenever an individual is dominated by the conviction that an object does not exist, the conviction renders him blind and deaf. If it can be truly said that miracles only appear to those who expect them, the converse is also true, and the preconceived idea that an object does not exist is a hindrance to seeing it. It may also be affirmed that every negation is far from being merely an inverted assertion, as it has been termed; it probably produces in a normal hearer a phenomenon of inhibition which is an attenuated form of systematic anæsthesia, just as the categorical assertion of a fact produces in a normal hearer a phenomenon of excitement which is an attenuated form of hallucination.

Nor, indeed, is a preconceived idea essential to the production of effects analogous to anæsthesia. The simple fact of attention, which consists in the concentration of the mind on a single point, has the result of increasing the intensity of that point so as to surround it with a zone of anæsthesia. Attention only increases the force of certain sensations in proportion as it attenuates others. One curious fact unites systematic anæsthesia with the negative effects of attention. We have seen that an invisible object, such as a square of red, may, when looked at with intentness, produce an after-image of the same form and of the complementary colour ; the object which was invisible in the case of direct vision becomes visible in this sort of consecutive vision. It is the same with objects at which we gaze without seeing them, since the attention is directed elsewhere. Physiologists say that, after looking at an object, an after-image is sometimes obtained in which details are discovered which had escaped the direct vision.

It is difficult to define the nature of systematic anæsthesia. We believe that the sensations produced by invisible objects penetrate to the sensory centre, since they produce after-images, and the cerebral seat of these images is no longer doubtful.* Moreover, the recognition of an invisible card among ten others which resemble it show that the subject's nervous centres had photographed all the details of the invisible object, as well as all the modifications to which it was subjected during the experiment. ·

It occurred to Richer to compare the mechanism of these phenomena with that of amnesia. We know that when memory does not respond to a spontaneous call, the material modifications which constitute its physical substratum still remain, since the amnesia may be transitory, and is curable. Something of the same kind occurs in anæsthesia by suggestion; the material modifications corresponding with the perception of the invisible object are produced, but they are not accompanied by consciousness. It is as if the subject, as soon as he perceived the invisible object, immediately forgot it. This is, however, only a comparison, not an explanation.

Nothing more definite can be said, except that systematic anæsthesia and other facts of the same class must be included in the great group of unconscious phenomena, in which they form a fresh class, with distinctive features of their own. Whatever be the part

---

* With respect to the cerebral seat of consecutive images, see Binet, *La psychologie du raisonnement,* p. 43.

assigned to consciousness, it is clear that the appearance of this phenomenon is connected with certain material conditions of the nervous centres; the conscious state implies different physiological conditions from the unconscious state. It may, therefore, be assumed, that suggested anæsthesia does not only destroy the phenomenon of consciousness, but that it modifies to a certain extent the concomitant nervous process. Indeed, we cannot understand how a suggestion should modify a psychical phenomenon without affecting the nervous process on which it is founded.

We have not yet had occasion to show the action of æsthesiogens on suggested anæsthesia; this action is curious, since the anæsthesia is destroyed. If, for instance, it has been suggested to an hypnotic subject that she does not see X——, who is standing before her, and a magnet is then applied to the back of her head, the anæsthesia presently disappears, and X—— again becomes visible. This experiment is the more curious, since the magnet also possesses the property of producing anæsthesia. Therefore the agent exerts two opposite effects, according to circumstances, and it may also be observed that when the action is continuous, these opposite effects are alternate, and this gives rise to consecutive oscillations.

Systematic anæsthesia is a comparatively simple phenomenon, accessible to observation and experiment, and for this reason it may serve as an introduction to the study of other phenomena of the same order which are much more complex.

Like hallucination and impulse, anæsthesia is a phenomenon which affects the peripheral parts of the intelligence, the senses and the movements. On the other hand, the complex phenomena connected with it are not external, and in some sense belong to interior, central psychology; they are seated in a region which for the most part eludes direct observation. We will only cite, as an instance of these complex phenomena, the failure of memory with respect to a letter, a word, or a whole language. This experiment was often performed by the early magnetizers, and it never failed to make a strong impression upon their audience. They requested one of those present to come upon the platform, and after making some passes over him, said abruptly in an imperious tone, "You have forgotten what your name is." The person addressed would make signs of denying the fact, would attempt to reply, and after wearying himself by opening his mouth in fruitless efforts, would finally confess that he no longer knew what his name was. The general astonishment can be imagined. This experimental amnesia appears to be allied with systematic anæsthesia; it is of the same order, with this difference, that it affects the images of the memory, instead of the external sensations and perceptions.

Some of our experiments confirm this idea. A few days after rendering F—— invisible, a somewhat curious fact was observed. The subject was able to see F——'s person, but did not recognize him, nor remember his name, although she had been acquainted with him for ten years. The suggested sensory anæsthesia had, without suggestion, spontaneously produced anæsthesia of the memory.

CHAPTER XII.

PARALYSIS BY SUGGESTION : MOTOR PARALYSIS.

MOTOR paralysis by means of suggestion forms one of the
most interesting and most carefully observed branches
of hypnotism.   We must first devote a few words to the
history of this form of paralysis, of which the discovery
is due to clinical science, and not to hypnotism.

It was in 1869 that Russell Reynolds first noted the
existence of motor and sensory disturbances, developed
under the influence of an idea.[*]   The motor disturbance
sometimes consists in spasms, in ataxic or inco-ordinated
movements, and more frequently in paralysis which
affects the upper limbs.   Erb gives to these symptoms
the name of imaginative paraplegia.[†]

The type of this paraplegia is afforded by Reynolds's
first observation, which concerned a young woman who
was affected by paraplegia under the following circumstances.   She lived alone with her father, who had
undergone a reverse of fortune, and who became paralytic in consequence of protracted anxiety.   She sup-

[*] Russell Reynolds, *Remarks on Paralysis and other disorders of
Motion and Sensation, dependent on Idea* (*Brit. Med. Journal*, pp. 378, 835,
vol. ii., 1869).

[†] Erb, *Paraplegie durch Einbildung* (*Handb. d. Krank. d. Nerven-
system*, p. 826, in Ziemssen, vol. xi. part ii., 1878).

ported the household by giving lessons, which involved long walks about the town. Influenced by the fatigue caused by so much walking, it occurred to her that she might herself become paralysed and that their situation would then be terrible. Haunted by this idea, she felt a growing weakness in her limbs, and after a while was quite unable to walk. The pathology of the affection was understood by Reynolds, who prescribed a purely moral treatment. He finally convinced his patient that she was able to walk, and in fact she resumed the practice.

Reynolds reports another remarkable case in which, although there was no real want of motor power, there was such a failure of motor co-ordination that walking became impossible. It was a case of this kind which led Charcot to study psychical paralysis, and he applied himself to show that the interpretation given by the English writer was legitimate. On this occasion Charcot again demonstrated the advantage to be derived from hypnotism in the experimental study of the phenomena which are spontaneously displayed, both in health and disease.

If the idea is impressed on a somnambulist that her right arm is paralysed, we see that this limb does in fact lose its motor power, and if the suggestion is made with that intent, the paralysis is maintained on awaking. Charcot has shown that this form of paralysis often displays objective characters which approximate to those of organic paralysis.

Moreover, as Bernheim has observed, many subjects who have been previously hypnotized may, without being hypnotized anew, display in the waking state an aptitude for the same suggestive phenomena; that is, paralysis by suggestion may be produced in some subjects, sensitive to hypnotism, when in the waking state.* Nor is this all. Bottey has confirmed Bernheim's researches by showing that in the case of some subjects who have never been hypnotized, paralysis may be produced by strongly impressing on them the idea that they are going to be paralyzed.† We thus come back by experiment to Reynolds's psychical paralysis.

Charcot, after repeating these various experiments, has shown that there is no breach between the somnambulist, and the subject liable to suggestion, but only a gradual transition, which enables us to understand psychical paralysis, and which demonstrates its reality. The experimenters of the Salpêtrière have aimed more especially at throwing into relief, according to the experimental method indicated above, the clinical characters of the paralysed limb. It is the more important to recognize these clinical characters, since they serve as a proof of the genuineness of the experiment, and make the nature of this paralysis by an idea in some degree intelligible.

We will now give an account of a case in which one of our subjects was paralysed by suggestion, taking care to insist on the material phenomena of which we have just spoken. The subject was first thrown into a lethargy by ocular pressure, and then into somnambulism by friction of the scalp. She was then told that her right arm was completely paralysed. It was necessary to insist on this assertion in a determined manner, since the

* *De la suggestion hypnotique*, p. 47.   1884.
† *Société de Biologie*, March 15, 1884.

15

subject offered resistance; she shook her head, moved her arm so as to show that she was able to do so, and replied to the experimenter, "I tell you I am not paralysed." The experimenter rejoined with unwearied repetition, "Your arm is paralysed: it is heavy; you cannot hold it up, it falls slackly to your side." In proportion to the repetition of these words, the subject moved her arm with increasing difficulty. Finally the paralysis was absolute, and the subject was altogether incapable of moving it.

This being the case, we said to her: "This paralysis will continue when you awake." We then awoke her by lightly breathing on her eyes. She was much surprised to find that her arm was paralysed, since her mind had retained no recollection of the suggestion made during somnambulism. She took up the pendent, paralysed arm with the other hand, and placed it on her lap.

It may seem surprising that when subjects awake in this manner to find themselves affected by a serious or revolting complaint, they show hardly any uneasiness. Unless a sensation of pain has also been suggested, they are rarely troubled by their condition, and they seem to be quite easy as to the issue of this affection. But it must be remembered that we have to do with hysterical patients, who endure with the greatest indifference all the sufferings which are the spontaneous result of their disease. For instance, they display no impatience if their limbs remain contracted for months together, and they do not deplore their lot as patients would do who are similarly affected in consequence of some organic lesion. This is a well-known feature, peculiar to hysteria.

When we examine the phenomena which are displayed by the paralysed limb, we shall first be struck by the complete destruction of motor power. The subject is incapable of performing the slightest movement. When desired to move her arm, she makes futile efforts and ineffectual contortions. It sometimes happens that when attempting to move the paralysed right arm, the subject performs involuntary and unconscious movements with the left arm. Besides being unable to raise the arm, the subject cannot hold it up, when it is raised by the hand of another; as soon as the arm is left to itself, it falls back like an inert mass. Its flaccidity is complete. This motor paralysis is generally accompanied by insensibility of the skin, an insensibility which may be profound, even if the subject was not anæsthetic before the experiment began. The arm may be pricked or pinched with impunity, and without producing the slightest reaction. If this proof is considered insufficient, recourse may be had to electricity, and intense currents may be sent through the paralysed arm without inducing the subject to complain; she remains perfectly passive.

With respect to the distribution of the anæsthesia, Charcot has recently ascertained that in some subjects anæsthesia is exactly coextensive with the paralysed region. If only the shoulder joint and its movement are paralysed, that region alone is affected by insensibility; the arm, the fore-arm, the wrist and fingers retain their normal sensibility. If the articulation of the shoulder and that of the elbow are paralysed, the anæsthesia goes lower, advancing as far as the middle of the fore-arm; if the wrist is also paralysed, the anæsthesia extends still further, and it finally affects the extremities when the fingers have also been paralysed.

Muscular sensibility is destroyed, as well as the sensibility to touch and to suffering; that is, the subject is unaware of the position occupied by the paralysed limb, and of the passive movements communicated to it. When the subject's eyes are closed, she is incapable of finding the hand of her paralysed arm with her sound hand; in order to do so, she must have recourse to the expedient of feeling for her shoulder, and passing her hand thence all along the arm to its extremity. When not allowed to use this means, the search occupies much time, and may even be indefinitely prolonged by withdrawing the paralysed hand from the sound hand which is seeking for it.

Conversely, if instead of suggesting the loss of motor power, the subject is told that throughout the limb there is a loss of cutaneous and sub-cutaneous sensibility, that her skin is insensible to contact, to pricks, etc., the association of symptoms produces more or less disorder in the motor functions. In some subjects a complete paralysis is produced; in others the effect is more superficial, and yet still more curious. The subject whose right arm is affected by anæsthesia, is unable, when her eyes are closed, to use this arm; when told to put her right hand on her forehead, she executes the gesture in question with her left hand. Sight is required in order to rectify this confusion between the two hands, and the subject makes no mistake when her eyes are open, and fixed upon the limb. We need not say more of this peculiarity, which is known to pathologists in other connections.

All the foregoing phenomena are purely subjective; there are others which are displayed by external signs.

The paralysed limb is cold, and the sensation of cold experienced by the subject may sometimes be verified by thermometrical observation. The motor signs must also be noted. Motor paralysis is accompanied by an exaggeration of the tendon reflex, which can be demonstrated by a very simple process of investigation. Striking on the tendons at the back of the neck, or of the wrist, suffices to produce shocks in the arm which do not occur in the normal state. This exaggerated reflex action is still more readily displayed in the leg, in which percussion of the patellar ligament produces a considerable shock (Charcot). This character connects paralysis by suggestion with organic paralysis. But, like the spontaneous paralysis of hysteria, paralysis by suggestion may present some variations in its symptomatic forms.

Richer and Gilles de la Tourette performed some interesting experiments with respect to the form of the muscular shock, by means of Marey's graphic method. They ascertained that during the period of paralysis the shock increases, and that it diminishes with the return of voluntary movement. In some cases, in addition to the increase in the height of the shock, they observed the broken and prolonged line of descent, which resembled an imperfect tetanisation. We know that when a contracture is produced during lethargy by the excitement of a nerve branch, or by kneading of the muscles, this lethargic contracture presents the curious property that it is relaxed and completely destroyed, when the excitement is applied to the muscles which are antagonistic to the contractured muscles. The contractures and paralysis produced during somnambulism or in the waking state do not display the same characteristic. If the subject is thrown into a lethargy, she maintains the contracture or paralysis which is given to her, and the excitement of the antagonist muscles produces no effect on this phenomenon. In order to make it disappear, recourse must be had to suggestion, by which it was produced.

So far, our study has been confined to the paralysed limb, but when a limb is paralysed by suggestion, an interesting fact takes place in the limb on the opposite side; its strength is increased so as to compensate to a certain extent for the paralysis of the other limb. In his normal state one of our subjects was requested to grasp a dynamometer, with the following result :—

| With the right hand | ... | ... | ... | ... 39 |
| With the left hand | ... | ... | ... | ... 27 |

He was then hypnotized, and his right arm was paralysed by suggestion. The pressure of the dynamometer afforded the following results :—

| With the right hand | ... | ... | ... | ... 0 |
| With the left hand | ... | ... | ... | ... 37 |

This result may be explained by saying that the inhibition effected on the right side by suggestion produced dynamogeny on the left. The simultaneous production of inhibition and of dynamogeny in symmetrical points has been repeatedly noted by Brown-Séquard in his experiments in vivisection. He writes as follows:—"The diminution and augmentation of the power and activity of the nervous system generally, if not always, co-exist. The same excitement of a point in the nervous system which diffuses itself so as to produce the inhibition of a property or activity in certain parts of the nervous centres, in certain nerves and muscles of one half of the body, also produces dynamogeny in the corresponding parts of the other half. This occurs when the exciting lesion is unilateral. For instance, the division of the sciatic nerves generally increases the excitability of the motor centres of the cerebral surface of the corresponding side, while it diminishes the excitability of the corresponding parts of the opposite side. Analogous, and generally stronger effects are observed after the transverse division of a lateral half of the spinal cord, and especially of the medulla oblongata, or of the pons varolii." *

These phenomena seem to show that the artificial modification produced in one hemisphere tends to produce a modification in the opposite direction in the other. There must therefore exist between the two hemispheres, or, as it has been said, between the two brains, not only a functional independence, but also, under conditions which are not yet ascertained, a compensating power of supply. It may be remembered that we met with similar facts during our study of hallucinations.

In the foregoing cases we have only spoken of the total paralysis which takes possession of the whole limb. By a different mode of suggestion the paralysis may be restricted to a group of muscles which are habitually associated in one movement. The subject may be told that she is unable to bend one finger. In this case the paralysis is not total but partial, and it is accompanied by some interesting facts which do not occur in total paralysis, of which the following is an instance.

* Brown-Séquard, *Recherches sur l'inhibition et la dynamogénie*, p. 25 Paris, 1882.

We suggested to a subject that she would be unable to bend her thumb in a way which was indicated to her. After a moment, when time had been given for effecting the suggested paralysis, we awoke the subject, who remembered nothing and had no suspicion of the paralysis. We desired her to make a great effort to bend her thumb, which she attempted to do, but the converse of what was ordered and intended took place, and instead of bending the thumb towards the palm of the hand, it was forcibly extended. The experiment was carried on by herself, and the thumb was contractured in its extended position; by degrees the index finger lost the power of bending, and this was also the case with the middle and third fingers, which were gradually extended, and slightly contractured in their extension. Thus it appears that when the subject desired to bend her thumb, she could only extend it, and the importance of this species of motor *quid proquo* is entitled to a passing notice. The experiment may be connected with that of systematic anæsthesia. We have seen that when the vision of a red square has been destroyed by suggestion, the fixed gaze at this square enables the subject to see the green complementary image. In the experiment we are now considering, the paralysis of a group of movements produced, when the subject wished to execute these movements, those which were antagonistic to them. Such a movement may be compared to the after-image, and the same relation may be said to exist between the antagonistic movements as between the complementary colours.

Of this we can cite another proof. It will be remembered that the application of the magnet, as well as suggestion, produces sensory paralysis, but with characteristic differences. Thus, under the influence of the magnet, the vision of a red cross is changed into the vision of a blank cross on a green ground. Suggestion, on the other hand, when it destroys the conscious vision of a red cross, enables the subject to see a green cross. The motor paralysis produced by the magnet differs in an analogous way from the motor paralysis produced by suggestion. If the subject is told that she can bend her thumb, and the flexion movement is then paralysed by the application of the magnet, it will be seen that the subject is also unable to extend her thumb; she can do nothing with it at all. If, on the contrary, suggestion has paralysed the flexion, the movement of extension is retained. We may say that paralysis by the magnet, which includes the two antagonistic movements, is comparable with the blank cross, in which paralysis of the two complementary colours occurs, and that paralysis by suggestion, which does not affect the antagonistic movement of extension, is comparable with the green cross, which preserves the complementary colour intact.

Secondary Symptoms. Aphasia.—The course of this brief description has enabled us to see that suggestion does not merely produce an isolated symptom in a subject, but a complete disease. It is, indeed, a remarkable fact, that when the suggested symptom is one of a nexus of symptoms, the subject of experiment shows a tendency to display the whole nexus. For instance, we impressed upon one of our hypnotized subjects the idea that when she awoke her right arm would be paralysed. We were much surprised to find that when she awoke, not only her right arm was paralysed, but she was unable to utter a word. Her intelligence was not affected, and she perfectly understood what was said, but the extremity of her tongue was drawn to the left side, and moved with difficulty. It was impossible for the subject to divine that this coincidence, interesting in more than one particular, could occur. This association is explained by the vicinity of the motor centres of the right arm, and of those for the muscles concerned with the production of articulate speech in the cortex of the left hemisphere of the cerebrum.[*]

So far we have been only concerned with the flaccid forms of paralysis. Contractured paralysis may be produced by the same process, since it is subject to the same laws as flaccid paralysis, and likewise causes an augmentation of muscular power in the corresponding limb. The special feature of suggested contractures consists in the possibility of impressing upon them the character of systematization which belongs to lethargic contractures. We have already observed that during lethargy the excitement of a nerve produces the contracture of the muscles in connection with it; the excitement of the ulnar nerve, where it passes behind the internal condyle, produces the well-known ulnar contracture. What is effected by mechanical excitement may also be effected by suggestion; when the idea is suggested to the subject of a pressure exerted at a level with her elbow, an ulnar contracture is produced which cannot be distinguished from that of lethargy. This experiment may be varied by describing a small circle on the lethargic subject's forearm, and by pressing the centre of this circle with the finger; a certain number of the muscles become contrac-

* Ch. Féré, *Les Hypnotiques,* etc.

tured.   The subject is then caused to pass from lethargy into somnambulism, and is told that on awaking she will feel a strong pressure on the centre of the small circle which has been described on her fore-arm.   When she awakes, the subject complains of pain, seated in the spot indicated by the suggestion, and her hand soon becomes contractured, and reproduces precisely the attitude it had assumed during lethargy.   This experiment shows that the suggested idea of an excitement, the image of a cutaneous excitement, may produce effects as intense and as precisely localized as an actual excitement.

This does not imply that lethargic contracture is a contracture produced by suggestion.   Suggestion and the physical impression constitute two parallel methods, and it would be very illogical to refer the method by excitement to the method by suggestion, because the latter is really derived from the former.

When once produced, paralysis by suggestion may be indefinitely prolonged.   We have seen a case in which it was maintained for twenty-four hours, and was unmodified by natural sleep.   The attempt to put an end to it provoked more resistance than if the paralysis had been recent, and there can be no doubt that without such intervention, it would have become more intense, and more difficult to cure.   The usual mode of destroying physical paralysis is to suggest the opposite idea of motor power; a simple assertion will not suffice, and it is necessary to insist, and to return to the charge, repeatedly telling the subject that she can move the limb if she chooses to do so.   Under the exciting influence of this suggestion, the subject attempts to raise the stiffened arm, the power of movement gradually returns to it, and the nervous circulation is restored, but some of the objective signs of paralysis, such as the exaggeration of the tendon reflex, remain for a little while longer. Another mode of curing motor paralysis is often quicker and more efficacious than suggestion; namely, to represent movement to the subject, either by actually performing such movements before him, or by impressing passive movements on the paralysed limb. The influence of such operations is well shown by the fact that, in the case of a healthy subject, the representation of a movement produced by one of the methods just indicated is calculated to increase the motor power.* It is a still more effectual process to induce the subject to move his sound limb, and then attempt to imitate these movements with the paralysed limb. In this way the subject carries on his motor education, at once by the muscular sense and by the sense of sight.

In concluding this clinical study of paralysis by suggestion, it should be added that the physical characters described above are not absolutely constant, and have not occurred in the subjects of several experiments. But their inconstancy does not affect their value and importance. All the foregoing experiments were performed on typical subjects, that is, on the hystero-epileptic subjects who display all the characters of profound hypnotism. These subjects have the advantage of displaying, with considerable exaggeration, symptoms which are rudimentary or altogether absent in normal subjects. From this point of view it may be said that the subjects of profound hypnotism constitute analytic cases which are eminently adapted for nosographical study.

* Ch. Féré, *Sensation et Mouvement*, etc.

II.

Systematic paralysis differs from total paralysis in its more complex character.   It consists in the loss of special and adapted movements.   The subject affected by it does not completely lose the use of his limb, but he is incapable of using it to perform a given act, and that act only.   Thus, the subject may be deprived of the power of performing the movements necessary in the action of sewing, drawing, writing, smoking, singing, playing on the piano, etc., while other movements are not affected. The authoritative assertion that the subject will, on awaking, be unable to write will, if repeated often enough, produce, by a mechanism which we do not yet understand, a paralysis of the power of writing, which is termed agraphia.

It is for the experimenter to choose the form which the systematic paralysis is to take ; it may be varied indefinitely, just as the form of hallucinations may be indefinitely varied.   Suggestion can reproduce all physiological phenomena, and the old magnetizers often took advantage of this fact.   They said to their subjects : " You cannot leave the circle I have drawn round you," and the subject, remained glued to the spot, in spite of all his efforts to leave the circle.   Or : " You cannot pronounce your own name," and the subject vainly opened his mouth, without being able to utter a syllable of his name.   Dr. Phillips, who in 1860 held public *séances* in hypnotism in Paris, on one occasion suggested to a person present, who was called Laverdant, that he would be unable to pronounce or write the two *a*'s in his name.

The subject tried in vain to write his name, and he traced the characters given in the accompanying facsimile (Fig. 15).

At first sight, systematic paralysis appears to have nothing in common with total paralysis. When a subject

Fig. 15.

has received the suggestion that she will on awaking be unable to write, no visible modification occurs in her arm. The paralysis remains in some sense latent, and is only revealed to the subject's consciousness at the decisive moment when she takes up a pen and tries to write. Up to that time her right arm appeared to be as perfectly free as the left, and displayed nothing in common with the flaccidity of total paralysis. Yet these two forms of paralysis only differ in degree. Total paralysis involves the loss of all kinds of movement, extension, flexion, rotation, abduction and adduction, etc., while in systematic paralysis, the loss does not apply to all movements, but only to those which are necessary for the performance of a given act.

To speak more precisely, when the subject is deprived of the power of performing a certain act, all the movements which have to do with that act are paralysed by suggestion. This is the plain fact. Suppose that suggestion has destroyed an act in which the movement of extending the index finger occurs. It may be asked whether the subject, who cannot perform the act as a whole, is incapable of extending the index in an isolated movement. In other words, is the systematic paralysis a loss of movement, or a loss of the power of co-ordinating certain movements with a view to an act ?

Experiment alone can reply to this question. Give to a subject a suggestion of agraphia, and when she awakes, examine her right hand. It is easily ascertained that the power of bending and extending the fingers remains, although these movements are largely employed in the act of writing, so that the individual movement is not lost, but the possibility of co-ordinating these movements so as to accomplish a given act. Another instance throws a still stronger light on this fact. Deprive a hypnotized subject of the power of writing the word *not*, and she will still be able to write at your request many other words, even those which contain the letters *n, o, t*, which proves that she has not lost the power of writing each of these letters separately, but only the power of combining them. Systematic paralysis therefore consists in a disturbance of motor co-ordination; it does not affect movements, but the association of movements; it disassociates the movements which were originally associated.

A useful comparison may be made between systematic paralysis and systematic anæsthesia, of which we have given a slight sketch above. There is a correspondence between these two phenomena, since one is in the series of facts of motor power, the other in the series of facts of sensation. There is paralysis in both cases, and in both the paralysis has a systematic character. The total paralysis of a limb corresponds with the total blindness of an eye, and the incapacity to perform a given act, and that act only, corresponds with the incapacity to perceive a given object, and that object only. We therefore decline to apply to systematic anæsthesia the name of *negative hallucination*, which appears to us to be singularly inappropriate, since we are not at all concerned with hallucination. To call systematic anæsthesia a negative hallucination is much the same as it would be to call systematic paralysis a negative motor impulse. We must set aside a vicious terminology which only serves to confuse our ideas.

The comparison we have just indicated might be carried further if our space permitted it. We can only point out one conclusion which may be deduced from it; that systematic anæsthesia, since it resembles paralysis, consists in great measure in a disturbance of the co-ordinating faculty.

We return to the study of systematic paralysis after this short digression. One character connects it with total paralysis : it is generally accompanied by a weakening of the motor power. If we turn to the former example, which is the simplest, we find that the subject whom we have caused to be affected by agraphia generally complains that his right hand feels somewhat heavy and inert. These subjective sensations are confirmed by direct examination. If the subject of agraphia is requested to grasp the dynamometer with his right hand, it is generally found that he exerts less than the normal pressure. This slight paresia has been pointed out by Pitres in a minute clinical observation, whence it appears that the artificial agraphia produced by suggestion presents this characteristic in common with spontaneous agraphia.

We have noted a second physical sign, which we also think important. Systematic paralysis, as well as total paralysis, produces a manifestation of dynamogeny in the symmetrical limb. When the right hand is affected by agraphia, the left hand becomes capable of registering on the dynamometer a higher degree than before. The loss on the one side is compensated by a gain on the other. This experiment may be carried further. The systematic· paralysis of one limb may not only produce in the other an increase of intensity in the muscular contraction, but also greater accuracy and perfection of movement. When a subject's right hand had been rendered agraphic by suggestion, she was on awaking requested to set down figures with the left hand. She consented, and the figures, which were reversed, as in a mirror, were almost irreproachable as far as the writing was concerned. The figures were all set down with one movement, and continuously, nor did the subject pause to consider. On another occasion we observed this subject's normal writing with the left hand, when the right hand was not agraphic. She then wrote with great difficulty; each figure cost her at least a half-minute's reflection, and moreover the result was somewhat defective.

The agraphia of the right arm consequently increases the co-ordinating power of moving the left arm in writing. This experiment may be explained by saying that, owing to the suggested agraphia, the faculty of writing, acquired by the right hand in consequence of a long apprenticeship, is transferred to the left hand.

These facts will remind the reader of the experiments in transference by the magnet which we have given

above. We saw that when the magnet was applied to a
subject who had received the suggestion of writing with
the right hand, the impulse was transferred from the
right to the left hand; the subject wrote backwards, as
in a mirror, with the left hand, and at the same time the
right hand became agraphic. The direct suggestion of
agraphia produces an analogous result, and this is explained by the fact that, whatever be the nature of the
excitement, whether suggestion or the application of the
magnet, the brain remains the same, and always re-acts
in accordance with the laws which govern it. This
reason also explains why Brown-Séquard produced the
same effects in dogs and guinea-pigs, by a twofold
organic lesion, which are produced in hysterical subjects
by transference.

We must in conclusion point out one more characteristic of systematic paralysis, which is also found in total
paralysis. The following observation was made by
Richer, who was not aware of its importance, although
he recorded it with the scrupulous conscientiousness of
an observer, and this makes it the more significant.*
" When X—— was in the somnambulant state, we told
her that she was unable to write. . . . As soon as she
awoke, we requested her to write her name. She took
up a pen with eagerness, but it had hardly touched the
paper, when she found it impossible to write a stroke,
however anxious she was to do so. It was interesting
to observe her gestures. At every effort which she made
to bend her fingers, they assumed movements of extension.

* *Op. cit.* p. 747. This observation was made on March 1, 1881, and
it agrees in all respects with experiments performed by us in December,
1883.

Her wrist was also extended, and her hand was raised. She tried to keep her right hand resting on the table with the aid of the left hand, but she was unable to restrain and regulate the contradictory movements which ensued from each attempt to write."

As we have already observed, the occurrence of antagonistic movements which accompanies the paralysis of certain movements, seems to us to resemble the production of the complementary colour which is observed in suggested achromatopsia. In both cases the paralysis of one function causes an exaggeration of the other, and it may be said that the same relation exists between the antagonistic movements as between the complementary colours.

When speaking of suggestions of acts, we insisted on the varied forms which the suggestion may assume. Suggestions of paralysis are susceptible of similar variations. Sometimes the subject is simply told that his arm is paralysed; sometimes it is suggested to him that he has forgotten how to move it, or again that he wishes not to move his arm, or the idea of an inability to move it is impressed upon him.

Most observers make use of several of these suggestions at once, cumulatively and without distinction, and it is noteworthy that such different processes afford identical results. From the psychological point of view there is a wide difference between the subject who does not move his arm because he is unable to do so, and the subject who does not move his arm because he does not wish to do so. But the clinical observation of these two kinds of paralysis shows that they present the same characteristics; it is, therefore, probable that in all these cases suggestion produces the same modification in the motor centres of the arm, in spite of its diverse forms.

We thus come to the conclusion that paralysis of the motor centre is the fundamental fact. This fact may be interpreted by the agent in different ways, and may be ascribed either to an incapacity to act, or to a determination not to act; but such interpretations are secondary, accessory, superadded phenomena, which do not form an integral part of the occurrence. The whole history of the will is comprised in two words, impulse and paralysis.

It may be asked what are the normal facts which may be compared with psychical paralysis by means of suggestion. Total paralysis, with its complete flaccidity, and its other strongly marked features, does not appear to correspond with anything in normal psychology; but this is not the case with systematic paralysis, since phenomena of inhibition occur in a sound person whenever his will effects an arrest of movement. Heidenhain observes that inhibition takes place when a man lowers his raised arm, and this is also the case when he refrains from the manifestation of violent anger or of fear. Ribot justly regards the will as at once a power of impulse and a power of restraint. As energetic a will is displayed in remaining impassive as in giving free vent to passion.

The action of æsthesiogens on paralysis by suggestion, merits examination. It is a remarkable fact that the transference which takes place in unilateral paralysis only occurs after a great convulsive discharge in the limb to which the paralysis is transferred, resembling a partial attack of epilepsy. When a bilateral paralysis is subjected to æsthesiogenic action, the corresponding impulse is substituted for it. The following example will suffice. When X—— was in a state of somnambulism, we suggested to her that she could no longer twirl her thumbs. She resisted the suggestion, saying that she could twirl them, and she did twirl them, but when the suggestion was repeated she stopped short. She was then awakened, and requested to make the movement in question; she tried to cross her hands, and was unable to do so. A small magnet was then placed at the back of her head, on the left side, without her being aware of it. After it had been there a few moments, she crossed her hands and twirled her thumbs. Presently she desisted, saying that she did not know how to do it, then resumed the movement, and continued it for five minutes without interruption, twirling her thumbs now in one direction, now in another. Meanwhile she talked of her friends among the patients, without thinking of what her fingers were about. We have already observed that the inverse effect is obtained by applying magnetic excitement to a subject who has received the suggestion of a motor impulse; in this case the movement is paralysed.

It should also be mentioned that in the case of many subjects a simple peripheral excitement, such as the compression of a limb, produces a similar inversion of the physiological state; the corresponding paralysis is substituted for the impulse, or the impulse for the paralysis. It is therefore probable that the magnet only acts as an unconscious peripheral excitement, of which the efficacy depends on the subject's physical condition.

What we have said of hallucinations, suggestions of acts, the paralysis of the senses, etc., also applies to motor paralysis. These comparatively simple and objective phenomena should serve as an introduction to the study of those which are more delicate and complex. We think that motor paralysis, the first order of systematic paralysis, naturally leads the observer to the study of the paralysis of the will, which has been termed *aboulia*. In order to make the meaning of this word intelligible, we may mention one of Bennett's patients, who was thirsty and requested the servant to bring a glass of water. When the glass was presented to her on a tray, the patient could not make up her mind to take it, although she wished to drink and her arm was not paralysed.

One of the present writers * has observed that when aboulia is produced by suggestion, it may become the source of delirious impressions which tend to become general. A subject who has been rendered incapable of seizing a given object will go on to say that it is not worth having, and will reject all the objects which resemble it.

We must now point out the analogy between aboulia and systematic paralysis; these two phenomena cannot be distinguished by any objective character. It can only be said that aboulia is an attenuated form of paralysis, but that it is at its maximum intensity equivalent to paralysis. Suppose that a subject affected by aboulia has at first experienced a certain reluctance to take up a pen and write, and gradually becomes incapable of doing so. At this point, his incapacity cannot be distinguished from a psychical paralysis of the movements employed in writing.

It may also be said that aboulia is a more complex

* Ch. Féré, *Impuissance et pessimisme* (*Revue Philosophique*, July, 1886).

state than systematic paralysis. To take, again, a case of agraphia, a subject affected by aboulia may be able to write anything except his signature, as was the case with a notary observed by Billod, while a subject affected by normal agraphia is unable to write anything at all. This difference is not, however, opposed to the fact that aboulia consists in a functional paralysis of one order of movements while all others are retained; it is in some sense an agraphia, but of a more systematic kind.

Whether, therefore, we have to do with total paralysis, with systematic paralysis, or with aboulia, the motor disturbance in correspondence with it is fundamentally the same, and we must recognize a cause which is analogous both in its nature and site. The analogy between these three phenomena can be illustrated by experiments in suggestion, performed on hypnotic subjects. We have seen that total paralysis and systematic paralysis produce in some cases an augmentation of force in the opposite limb. This is also the case in aboulia.

Suppose that it is suggested to a subject that he will be unable, however much he wishes it, to open a table drawer with his right hand, in order to take out something which is there; if, after having produced this unilateral aboulia, a dynamometer is placed in the subject's hands, it will be ascertained that the muscular force of the right arm has diminished, while that of the left arm has increased.

We can judge from this example of the value of a method which takes phenomena in a series. Aboulia is a complex pathological fact which cannot be grasped at once, and without preparation; we might as well begin geometry with the study of curves. Aboulia can only be understood after we have studied the simpler phenomena, which are more easy to observe and analyse. We consider that paralysis by suggestion presents the elementary phenomena which should serve as a basis and introduction to the study of aboulia.

In short, the great psychological conclusion which may be drawn from all the disturbances of motor power, is that these disturbances are directly caused by functional modifications of the motor centres: these are the true causes of motor paralysis, whether total, systematic, or phenomena of aboulia. It may even be said that in a case of total paralysis by suggestion, the effects observed are the same as when a knife has destroyed the motor centre in correspondence with the paralysed limb. The mode in which the subject interprets the motor disturbance by which he is affected is an altogether secondary matter. It is unimportant whether the motive assigned by the subject for his impotence is that he cannot, or that he will not, or again—as it is said by some subjects affected by aboulia—that he is not able to exert his will, or that he does not understand. If these assertions were taken literally, we might be inclined to regard these disturbances of the motor power as distinct phenomena, instead of recognizing the fact that they are included in the same order. When the subject says that he cannot move his limb, we shall see in this inability a simple motor paralysis. When the subject asserts that he really desires and yet is unable to resolve on passing through a door, a lesion of the will may be diagnosed. Finally, when the subject says he has forgotten how to make the movements employed in writing, such agraphia must be ascribed to a lesion of the motor memory. The logical consequence of this method would be to ignore the fundamental unity of these three motor facts, which are, as we must repeat, only variations of the same state, that of motor paralysis.

Our object in this study of the facts of suggestion is to show the importance of hypnotism as a psychological study. Its importance was advocated many years ago, but the cause of animal magnetism was so much compromised by erroneous methods, that people were afraid of entering upon such questions. After Braid had demonstrated the reality of a nervous state, produced by looking fixedly at a brilliant object, together with the possibility of producing, by verbal suggestion, many psychical phenomena in the hypnotized subject, it might have been supposed that psychologists would at length have begun to study these facts, in which they were so directly interested. This was, however, by no means the case. With the exception of a few isolated attempts, these fruitful studies were for the most part neglected, doubtless from the dread of compromising the inquirer. It is a matter of regret that the English psychologists of the associationist school—Stuart Mill, Bain, and Spencer —who, although not experimenters, strictly so called, yet have always evinced the greatest respect for experiment, never thought of availing themselves of the precious documents contained in the writings of their fellow-countrymen. It is strange that they did not understand that these afforded the finest illustrations of the general law of the association of ideas which justly appeared to them to be so important.

The indifference displayed by psychologists towards
16

hypnotism was such a well-known and established fact, that it was supposed to have nothing to do with psychology, and in 1871 Mathias Duval expressed the current opinion, when he wrote the remarkable article on hypnotism in which he asked, not without irony, "Where are the discoveries of hypnotism? Where are its analyses? Where are the results of this new experimental philosophy?"

A revolution has taken place since that epoch. Profound hypnotism, studied with so much precision by Charcot, has triumphed over the general indifference which stifles inquiry. It is now generally admitted that hypnotism constitutes one method of psychological research, which offers the twofold advantage of enlarging and isolating the states of consciousness.

It is the more important to insist on this fact, since we think that hypnotism is adapted to fill up a breach. It is some years since a certain set of writers endeavoured to establish an experimental psychology in France, in opposition to the classic psychology which is still supreme in all the universities. But up to this time we have been unable to note any marked distinction between the new and old schools. In reply to the inquiry into its characteristics, we are told in the first place of a dislike to metaphysics. But this is only a name, and it is hard to say where metaphysics begin, and positive science ends. Again, it has been said that the new psychology is experimental. But in order to be experimental it is necessary to perform experiments, and we must ask where these are to be found. They are few in number, and chiefly consist of observations relative to the measure of sensations, the time of reaction, etc. It seems to us that hypnotism, in association with the clinical observation of mental and nervous diseases, would afford to the new school the method of which it is in search, and would furnish conclusions founded on experiment.

CHAPTER XIII.

THE APPLICATION OF HYPNOTISM TO THERAPEUTICS AND EDUCATION.

I.

WHAT we have said of hypnotism, and particularly of suggestion, may lead the reader to understand the virtue of medicine for the imagination, of which the importance has already been intimated by earlier writers. Deslon asked why, if medicine for the imagination was the most effective, it should not be employed.

We must be permitted to dwell for a moment on this medicine for the imagination, which is entitled to the name of suggestive therapeutics. The process is as follows. Influenced by a persistent idea, suggested by external circumstances, a paralysis is developed. The physician makes use of his authority to suggest the idea of an inevitable, incontestable cure, and the paralysis is cured accordingly. This cure, as well as the development of functional disturbance, was directly effected by an idea. An idea may, therefore, be, according to circumstances, a pathogenic and a therapeutic agent. This notion is not new, but since it was misinterpreted, it has remained unfruitful.*

* Ch. Féré, *La médicine d'imagination* (*Progrès Médical*, p. 309, 1884, pp. 717, 741, 760, 1886).

Diseases have been termed imaginary, or diseases caused by the imagination, and this confusion of terms has confirmed the confusion of ideas. We have, however, just shown, especially by means of the facts which relate to paralysis by suggestion, that diseases caused by the imagination—that is, produced by a fixed idea—are real diseases, and, at any rate in some cases, display undisputed objective symptoms.

Since the existence of real diseases, produced by means of the imagination, is proved, it is thereby proved that imaginary diseases do not and cannot exist; by this we mean purely fictitious diseases, since as soon as the subject has accepted the fixed idea that he is affected by any functional disturbance, such a disturbance is in some degree developed. It should be added that these diseases, produced by means of the imagination, are not merely influenced by a local disturbance; the subject who allows himself to be dominated by this idea of disease must be peculiarly excitable and open to suggestion; he must be endowed with a condition of congenital psychical weakness which is frequently found in conjunction with more or less strongly marked neuropathic manifestations, or with physical malformations. As Lasègue observed, not every one who pleases can be hypochondriac.

This distinction throws light on the therapeutics of diseases produced by means of the imagination, or suggested diseases.

When one of these victims to hypochondria, anæmic and emaciated, who are usually called *malades imaginaires*, has recourse to medicine, on the plea of suffering pain or some other subjective disturbance, he is usually told that it is of no importance, that he is rather fanciful and should think less about his health, and some anodyne is carelessly prescribed. The patient, who is really suffering from the pain he has suggested to himself, feels convinced that his malady is not known, and that nothing can be done for him. The idea that his complaint is incurable becomes intense in proportion to his high opinion of his physician's skill, and thus the patient, who was suffering from the chronic affection suggested by his imagination, often goes away incurable.

Those who undertake miraculous cures act very differently. They do not deny the existence of the disease, but they assert that it may be cured by supernatural power. They act by means of suggestion, and by gradually inculcating the idea that the disease is curable, until the subject accepts it. The cure is sometimes effected by the suggestion, and when it is said to be by saving faith, the expression used is rigorously scientific. These miracles should no longer be denied, but we should understand their genesis and learn to imitate them.

There are, therefore, no imaginary diseases, but there are diseases due to the imagination, and accompanied by real functional disturbances. Such disturbances may be developed under the influence of spontaneous, accidental, or deliberate suggestion, and they may be cured under the influence of another suggestion of equal intensity working in an inverse direction. The moral treatment ought not, therefore, to consist in denying the existence of the disease, but in asserting that it is susceptible of cure, that the cure has actually begun, and will soon be completed.

When a believer associates the Deity with his idea

of cure, he is accustomed to expect it to be sudden and complete, as the result of a definite religious manifestation; and this, in fact, often occurs. We had a well-known instance at the Salpêtrière, when a woman of the name of Etcheverry was, after her devotions in the month of May, suddenly cured of an hemiplegia and contracture, by which she had been affected for seven years. Only a slight weakness of the side remained, which disappeared in a few days, and which could be explained by the prolonged inaction of the muscles. This may be termed an experimental miracle, since the physicians had prepared for it beforehand, having for a long time previously suggested to the subject that she would be cured when a certain religious ceremony took place, and it is a miracle which explains the numerous cures by the laying-on of hands which are recorded in the Bible. If we do not go further back than the last century, suggestion explains the cures by Greatrakes, the exorcisms by Gassner, Mesmer's successes, and the miracles performed at the tomb of the deacon Paris in the cemetery of Saint Médard; and in our day, in the famous caves on the slopes of the Pyrenees.

The resources of the physician, who does not profess to be a thaumaturgist, are more scanty. When he is consulted by a patient whose disease has a psychical origin, he is unable, unless in some exceptional circumstances, to inspire confidence in remedies which are not more or less gradual, but whatever they are, he must prescribe with firmness and authority. It is a well-known fact that the hydropathic treatment of some forms of hysteria has afforded more speedy results than other modes of treatment, merely from the fact that suggestion has been employed at the same time. This remark also applies to massage, etc., under analogous circumstances.

In many cases suggestion may become a valuable therapeutic agent. In addition to the paralysis and spasms which are of psychical origin, it has a great influence on nervous or hysterical anorexia, and on the anæmic disturbance which is generally developed on an hysterical soil (A. Voisin, Séglas, Lombroso, Dufour, etc.). Some weight must be given to several facts of this nature, reported by Braid, Charpignon, Liébault, Bernheim, Beaunis, etc.* It is, therefore, useful in such cases to inquire into the most favourable conditions of suggestion, to ascertain whether the subject is susceptible to hypnotism, or peculiarly sensitive to any mode of suggestion which is employed with confidence and authority.

It should be observed that a neuropathic state does not occur suddenly, and is not created by the person affected by it. It is generally the result of a progressive and accumulated hereditary degeneration. The subject under treatment does not essentially differ from the rest of his family, who usually suffer to some extent from the same evil; the nervous patient lives in a nervous environment. If suggestion has taken part in the development of the affection in question, the moral treatment may be ineffectual because the pathogenic idea continues to be cultivated in this morbid environment. Such treatment will only have a chance of success when, as a measure of moral hygiene, we isolate the

* Charpignon, *Étude sur la médicine animique et vitaliste* (1864); Liébault, *Du sommeil et des états analogues* (1864); Bernheim, *De la suggestion et de ses applications à la thérapeutique* (1886).

patient, and this is still more necessary in the so-called epidemic phenomena of suggestion.

We are particularly anxious to call attention to the effect of moral treatment, and to the part taken in it by suggestion. This is no new thing; when the so-called fulminating pills are administered, suggestion is employed in the pilular form, and when pure water is injected under the skin, suggestion takes a hypodermic form. This medicine for the imagination is particularly to be recommended in that category of diseases which are of definite psychical origin.

This is not the place for insisting on the special indications of suggestion in therapeutics. The study just made is enough to show to what extent it may act on motor, sensory or psychical phenomena, and consequently how it may be usefully employed in the treatment of the dynamic disturbances which are due to the influence of a psychical action, of a moral shock, or even of a peripheral excitement. Its effect cannot any longer be disputed. It is, however, still difficult to give a rigorously scientific account of the results obtained, since few observations have as yet been published, and in some of these it is impossible to find an objective characteristic of hypnosis. Others, again, are incomplete, or published by incompetent persons, whose descriptions do not carry with them a conviction of the reality of the morbid state in question. Finally, precisely on account of the nature of its action, which is exclusively exerted on diseases in which there is no definite material lesion, and which are, therefore, purely dynamic, suggestion only cures affections which are capable of spontaneous modification, or which are influenced by various external agents. At present, there-

fore, it is difficult to establish the real value of this mode
of treatment, although less difficult than in the case of
many remedies in general use.   It can only be said that
it is founded on accurate notions of mental physiology,
and consequently on a rational basis.

Medicine for the imagination is distinct from hypnotic
therapeutics, in which the artificial sleep is itself the
curative agent, in whatever way it may have been produced.   These two therapeutical processes, artificial sleep
and suggestion, have sometimes been erroneously confounded.*   They are far from being of equal value.

The hypnotic sleep often exerts a merely suspensive
and momentary action on functional disturbances, such
as neuralgia, contractures, etc.; but, taken by itself, it
rarely effects the complete disappearance of these phenomena, unless they are of an essentially fugitive nature.
It should also be noted that in many cases hypnotic
sleep, like other forms of artificial sleep produced by
chloroform, morphia, etc., may cause neuropathic affections which the subject has not experienced before.
Many hysterical patients were attacked by convulsions,
when assembled round Mesmer's *baquet*, and magnetizers
have often produced contractures of a cataleptic character.
This fact must not be ignored, and it shows that the
employment of hypnotism as a therapeutic agent should
not be undertaken rashly.

When, however, we have to do with strongly marked
cases of hysteria, in which the convulsive attacks are
intense, so that the artificial sleep can only produce
affections which existed previously, it has been ascer-

---

* Grasset, *Du sommeil provoqué comme agent thérapeutique* (or therapeutic suggestion, *Semaine médicale*, p. 205, 1886).

tained that the number and intensity of these attacks may be greatly diminished by hypnotic treatment. Many of the hysterical patients of the Salpêtrière, who have been admitted on account of their attacks, enjoy a respite from them whenever they are thrown into the hypnotic sleep, even when it is not accompanied by suggestion.

It should also be observed that the nervous sleep is something more than a simple hypnotic action. The subject is aware, when hypnotized by a magnetizer, that the object of his treatment is therapeutical, and the artificial sleep may in some cases be regarded as pertaining to the medicine of the imagination. Suggestion is present, whether it be the work of the patient or of his physician.

II.

The employment of suggestion in education is probably as ancient as the art of teaching, and interesting remarks on the subject may be found in many educational works. Fechtersleben, in his *Hygiène de l'âme*, insists on the benefit of convincing children that they have a gift for some branch of study, in order to develop their capacity. Gratiolet remarks that certain gestures and attitudes will develop correlative tendencies in children.[*] But it would be a serious mistake to subject children of normal constitution to the regular practice of suggestion; there would be a great risk of making them into automata, which is by no means the end of education. It is more easy to defend the application of hypnotic suggestion to vicious children. It is probable that it might succeed,

---

[*] Leuret and Gratiolet, *Anatomie comparée du système nerveux*, vol. ii. p. 630.

but it would be difficult to prove this by indisputable facts, for it is certain that some of the vicious children who escape from premature insanity or from a progress in vice, pass by spontaneous evolution into a psychical state which is almost normal. On the other hand, some of the vagrant children who have been confined in asylums are found under the same conditions as those whose penal sentence confirms the motives for avoiding vice. In these cases hypnotic suggestion only plays the part of penitentiary suggestion, and its utility is doubtful.

The efficacy of suggestion by teachers may, as we believe, be shown by the possibility of modifying certain instincts by suggestion in the case of animals. One of the present writers has repeatedly witnessed a curious practice employed by a farmer's wife in the district of Caux. When a hen has laid a certain number of eggs in a nest of her own selection, and has begun to sit, if there is any reason for transferring her to some other nest, the hen's head is put under her wing, and she is swung to and fro until she is put to sleep. This is soon done, and she is then placed in the nest designed for her; when she awakes, she has no recollection of her own nest, and readily adopts the strange eggs. By means of this process, hens may sometimes be made to sit which had shown a previous disinclination to do so. This modification of instinct by suggestion seems to show that the educational use of suggestion is not so absurd as some authors assert it to be.

CHAPTER XIV.

HYPNOTISM AND RESPONSIBILITY.

" Non omnes dormiunt qui clausos habent oculos."

SINCE the past history of hypnotism verged upon the marvellous, it had the privilege of exciting the curiosity, not only of learned men, but of people in general. Exhibitions with which science had nothing to do made the public acquainted with a certain number of phenomena of which a criminal use might be made; and hypnotic sleep and suggestion have played a part in several judicial dramas.* One of the present writers was therefore justified in being the first to call attention to the study of hypnotism from the medico-legal point of view, in a paper which only treated of profound hypnotism, characterised by symptoms of a physical kind.† Liégeois

---

* Tardieu, *Étude médico-legale sur les attentats aux mœurs*, pp. 88, *et seq.* (Paris, 1878); Brouardel, *Accusation de viol accompli pendant le sommeil hypnotique (Annales d'hygiène et de médicine légale*, January, 1879); Motet, *Annales médico-psychologiques*, p. 468 (1881).

† Ch. Féré, *Les hypnotiques hystériques, considérées comme sujets d'expérience en médicine mentale : illusions, hallucinations, impulsions irrésistibles provoqués ; leur importance au point de vue médico-legale (Société médico-psychologique*, May, 1883). J. Charpignon (*Rapports du magnétisme avec la jurisprudence et la médicine légale*, 1860) has been chiefly concerned with an inquiry whether the practice of magnetism does not constitute the criminal offence of an illegal practice of medicine.

has subsequently communicated to the Academy of the Moral and Political Sciences a paper on the same subject, regarded from a somewhat different point of view, and this has given rise to lively discussions.* We think it may be profitable to consider this subject, which is indeed entitled to further development.

Most of the writers on this question have been chiefly occupied in throwing light on the possibility of accomplishing criminal acts by means of hypnotism, but they have not considered the question of proof. They have not asked under what conditions judges would admit the reality of the facts of hypnotism brought before them. They have not understood that in a medico-legal study, the demonstration of the hypnotic state is the first question, and the most important of all; the others are comparatively unimportant, since if the hypnotism is not proved, all the consequences which may be drawn from it become illusory. It need hardly be added that a scientific demonstration of hypnotism can only be made by means of objective and material signs. Several observers have accepted as proofs the honesty and good faith of their subjects, but these words do not involve any objective sign. Moral proofs must always remain personal to those who appeal to them, and cannot be taken into account in a medico-legal study. We cannot hope to convince judges of the reality of a state in which all the phenomena may be simulated. To accept the fact hypnotism on the ground of moral proofs would be to open the door to innumerable abuses of the most serious character.

* Liégeois, *De la suggestion hypnotique dans ses rapports avec le droit civil et le droit criminel* (*Ac. des sc. m. et p.*, April and May, 1884).

The medico-legal question may be briefly stated in the following terms. An individual appeals to the law, asserting that he has been the victim of some violence, or of a suggestion, after he had been thrown into the hypnotic state. It may be admitted that the assertion is probable, if it is proved experimentally that he is susceptible to hypnotism, and that he displays a certain number of objective, characteristical phenomena, but this proof can only be obtained if he voluntarily submits to be subjected to experiment. Again, an individual accused of a crime or an offence, may plead that he acted under the influence of an impulse suggested during the hypnotic sleep, and in this case also it is necessary to obtain a material proof that the subject is susceptible to hypnotism. As a general rule, whenever an accused person pleads hypnotism, the fact should be proved by subjecting him to experimental observation.

Another situation may occur. A witness may be suspected of making a deposition dictated by hypnotic suggestion. If the fact of suggestion is established by material proof, the fact that he has borne false witness will also be demonstrated. If the material fact cannot be established, the difficulty is almost insuperable, since an individual cannot be constrained to submit to hypnotization, any more than to take chloroform or haschish.

The conditions which enable an expert to affirm that a person is susceptible to hypnotism are as follows. The hypnotized subject must display its physical phenomena, and must be a subject of profound hypnotism. Profound hypnotism may occur either in a perfect or imperfect form ; that is, some of the established phenomena may be absent in a given subject without affecting its general aspect, if a sufficient number of characteristic phenomena remain.

In slight hypnotism, in the states described as fascination, magnetic sleep, etc., subjects appear to be peculiarly liable to suggestion. It is possible to develop in them catalepsoid states, muscular rigidity, fixed attitudes, paralysis, anæsthesia, various hallucinations and impulses; but not the special and characteristic states described above under the names of catalepsy, lethargy, and artificial somnambulism. These subjects only display a few physical phenomena which have not yet been the object of a regular nosographic study. The strict attention with which the facts should be examined, must be redoubled in such cases, since the only criterion is afforded by physical phenomena. Until we know more of the subject, a person who displays none of the physical characteristics of hypnotism cannot plead it as a justification. It is practically impossible to define in any other way the limits of normal suggestibility.

After having shown how the expert may satisfy himself that a subject is susceptible to hypnotism, we must consider the special circumstances in which it is possible to admit the probability of hypnotization.

The hypnotic sleep, which is produced with so much difficulty and delay in fresh subjects, occurs with alarming rapidity in those who have been long under treatment. Some of our patients are hypnotized at once by a single abrupt gesture, and this may be effected in all places alike, and at any hour of the day. If we meet one of these subjects crossing the courtyard, an exclamation or abrupt gesture will cause her to stop short and become motionless in catalepsy. She may be as instantaneously awakened by breathing on her forehead or eyes. The hypnotic sleep may, therefore, be produced and brought to an end in an extremely short time, we might even say during the passage from one door to another. This is a somewhat important fact from the medico-legal point of view. And again, a suggestion may be given in a sleep of very short duration. We have observed that in the course of fifteen seconds we could throw a subject into a lethargy, then into somnambulism, suggest an act, and then awake him. · It is, therefore, possible that an individual might make use of the fifteen seconds in which he found himself alone with a susceptible subject to inculcate an idea, an hallucination, or an impulse. We must not rely on a question of time in maintaining the impossibility of such a fact, since the time required to hypnotize and suggest to an habitual subject of experiment is so extremely short.

Experience also shows that we must not accept the subject's assertions in estimating the duration of hypnotic sleep. The subject is unable to measure the length of time she has slept, and if she attempts to do so she makes the gravest mistakes. For instance, one of our subjects whom we had hypnotized for a period which did not exceed twenty seconds, believed that she had slept an hour, and in other cases there was equal miscalculation. The hypnotic subject has no landmarks by which to measure the void which this sleep produces in the normal course of life. We must not, therefore, deny the reality of an hypnotic suggestion, merely because the experimenter had only a minute's contact with the subject, although she asserts that she has slept for hours.

We unfortunately possess but few documents bearing on the question how far the subject is aware that he has been hypnotized. Some of those on whom we have been performing experiments for a whole morning, do not know how often they have been hypnotized and awakened ; but they have a general knowledge of having been subjected to hypnotism. They know this from an impression of cold, a shivering which often lasts long after they are awake. But this sign is not of much value, since it may not only be absent, but destroyed by suggestion, and it is less marked in proportion to the shortness of the sleep.

In cases of profound hypnotism there is often an oblivion of what occurred during the hypnotic sleep. This oblivion is complete when the experimenter has taken care to tell the hypnotized subject that he will remember absolutely nothing. The oblivion is also rendered more profound when the subject has not been directly recalled to the waking state, but has passed from somnambulism into lethargy, from lethargy again into somnambulism, and thence into the waking state. On the other hand, the amnesia is often only partial when the subject is awakened immediately after the occurrence of a given fact; a more or less vivid recollection of it still remains. The hypnotic subject seems to be in the same situation as a man who awakes from normal sleep; he has a vague recollection of what he has seen, or of what has been said to him during sleep, and he thinks that he has been dreaming. Finally, the events which occurred during hypnosis recur to his mind with great force, when they are recalled by some external object or circumstance.

Thus it appears that it is impossible to lay down an absolute rule as to oblivion on awaking; there is, in fact, every variety of case, from the most profound oblivion to the most lucid recollection, and these are all entitled to careful consideration from the medico-legal point of view.

A suggested amnesia is the most important of these situations. It must always be borne in mind that a suggestion will destroy the subject's recollection of all which has occurred to her during hypnosis. This want of memory, which may either occur spontaneously or be artificially produced, is possible even when the subject has experienced a shock, of which the effects are painful or more or less enduring. In the course of an experiment, one of our subjects, who was in a state of lethargy, fell down and knocked her head violently against the floor. She was not awakened by this excitement, nor for some time afterwards, and she was then awakened by breathing on her face. On coming to herself, the subject was astonished by the pain in her head; she had the sensation of a violent blow or shock, and could not understand whence it came. We are, therefore, justified in the assertion that a subject of profound hypnotism may undergo all sorts of violence without retaining any recollection or consciousness of it, unless the violence has produced permanent lesions, such as the attrition of the tissues resulting from a violent shock, etc. We even think it possible that a subject might be violated in the hypnotic state, in which she would be unable to offer any resistance.

We turn from the subject who has forgotten every-thing, to the subject who asserts that she remembers everything, in order to consider whether her story is to be believed. It is a serious question, and admits of many hypotheses.

In the first place, the subject may be perfectly honest, and yet the victim of an illusion. When a subject finds on awaking that she is suffering from a wound, or from some serious or unpleasant affection, she is apt to look for an explanation, and sometimes she invents an explanation for herself. Sometimes, again, she accepts it from a third person, but in all cases she ends by suggesting to herself that she saw things occur as she has explained them; in other words, the explanation has led to an hallucination of memory. Thus a subject who has, during hypnosis, received a blow from a third person, may explain her injury by the supposition that she fell down, and she will maintain the reality of this imaginary fall with the strongest conviction. The medical jurist must be on his guard against the remarks and explana-tions adopted by the subject to account for the accidents which happen to her, and her assertions should not be accepted without confirmation.

The subject may err from another cause, the sugges-tion of the experimenter, who has impressed upon her a recollection which is false. It is impossible for the expert to steer clearly amid all these phenomena, and to make a categorical declaration as to the way the thing occurred.

Finally, the subject who in the waking state remem-bers what occurred during hypnosis, may simulate. This danger of simulation is always present in a legal case, whatever be the physical state of the subject. Even in a case of profound hypnotism, we must not blindly accept whatever is related by the subject. The testimony may be recorded, and taken for what it is worth, in conjunction with other facts, but this is not the business of the expert.

We have hitherto regarded the subject in a state of repose, and we must now consider him in a state of activity, influenced by suggestions or excitements. We will begin with the study of hallucinations. The subject may, for instance, be induced to mistake the identity of a person, or to accept the presence of one who is really absent, and to recognize his features, voice, etc. The possible consequences of this illusion or hallucination are evident. If an unlawful or criminal act should be committed on the subject, or in her presence, an accusation might be made against an innocent person, and it would be maintained with the deepest conviction. The illusion or hallucination might apply to the act itself, and would lead to analogous consequences.

Some writers have lately returned to this question, of which we long ago noted the importance. They devised some dramatic experiments, to illustrate the criminal use which might be made of hypnotic hallucination, but we do not think it necessary to reproduce them. It seems to us more important to inquire under what conditions these facts of retrospective hallucination can be adduced in a court of justice. As we have already said, the medico-legal question of hypnotism is reduced to a question of diagnosis, and it may be added that whatever is not diagnostic in the legal questions which affect hypnotic subjects is not within the province of the physician. The expert has not to decide upon the reality of a fact of suggestion, but on its possibility; and in order to do this, he must establish by experiment the fact that the phenomena under dispute can be reproduced in a given subject, by means of an hypnotic suggestion. We are, therefore, concerned to know what are the objective signs which demonstrate the hallucinations produced in a given subject to be genuine. We propose to dwell chiefly on hallucinations of vision, since they are the most easy to examine, and we can deduce from them the sincerity of the hallucinations of the other senses.

Hallucinations of vision, whether induced by verbal suggestion, or by any other process, are chiefly characterized by their power of duplication, either when a prism is placed before the eyes, or when a mechanical deviation is effected. The hallucinatory object may be enlarged or diminished in size by a lens, and may be reflected and rendered symmetrical by a mirror. If a coloured object is in question, it may give rise to a subjective sensation of the complementary colour, and if the hallucination is unilateral, it may be transferred to the other side by the action of the magnet. Finally, if the imaginary object is brought nearer, or withdrawn, there is a corresponding dilation and contraction of the pupil. These movements to accommodate the sight occur spontaneously in very few cases, and only under known conditions.

It should be added that in the case of some subjects, the general sensibility of the eye is profoundly modified during the period of visual hallucination; there is, in fact, speaking generally, insensibility of the conjunctiva and of the cornea, as well as of the region of the pupil.

These may, in the case of most subjects, be touched with a foreign body, without producing any reflex action of the eyelid. In P——, however, when a visual hallucination has been developed, the sensibility of the external membranes of the eye is normal.

All these signs enable the expert to know whether a subject can or cannot receive an hallucination by suggestion, but they by no means establish the fact that the hallucination has actually occurred. This is a distinct question, which it is the part of the legal examination to elucidate.

Systematic anæsthesia serves as a counterpart to hallucination, and must be considered in conjunction with it. It may be suggested to a subject before whom a crime or offence is to be committed, that he is unable to see a given person. The possibility of making such a suggestion would enable a criminal to get rid of a troublesome witness, and the subject might then declare in a court of justice, in all good faith, that he had seen, heard, and felt nothing.

We need only remind our readers of the numerous proofs of systematic anæsthesia. The first of these is the Chinese gong. A subject susceptible to hypnotism who instantly becomes cataleptic at the sound of the gong, is no longer affected by catalepsy when it is sounded after his perception of the instrument has been destroyed by suggestion. A second proof is derived from the complementary colours. When it has been suggested to the subject that he cannot see the colour red, his fixed gaze at an invisible square of red produces after a while a consecutive green image. Finally, the action of the magnet impresses a special character on systematic anæsthesia. The expert may ascertain from these signs whether a given subject, placed in a state of hypnotism, is capable of affording authentic instances of anæsthesia.

It is possible to suggest to a subject in a state of somnambulism fixed ideas, irresistible impulses, which he will obey on awaking with mathematical precision. The subject may be induced to write down promises, recognitions of debt, admissions and confessions, by which he may be grievously wronged. If arms are given to him, he may also be induced to commit any crime which is prompted by the experimenter. We could cite several acts, to say the least unseemly, committed by hysterical patients, which were crimes in miniature, performed by an unconscious subject, and instigated by one who was really guilty, and who remained unknown. At the Salpêtrière a paper-knife has often been placed in the hands of an hypnotic subject, who is told that it is a dagger, with which she is ordered to murder one of the persons present. On awaking, the patient hovers round her victim, and suddenly strikes him with such violence that I think it well to refrain from such experiments. It has also been suggested to the subject to steal various objects, such as photographs, etc.

These facts show that the hypnotic subject may become the instrument of a terrible crime, the more terrible since, immediately after the act is accomplished, all may be forgotten—the crime, the impulse, and its instigator.

Some of the more dangerous characteristics of these suggested acts should be noted. These impulses may give rise to crimes or offences of which the nature is infinitely varied, but which retain the almost constant character of a conscious, irresistible impulse; that is, although the subject is quite himself, and conscious of his identity, he cannot resist the force which impels him to perform an act which he would under other circumstances condemn. Hurried on by this irresistible force, the subject feels none of the doubts and hesitations of a criminal who acts spontaneously; he behaves with a tranquillity and security which would in such a case ensure the success of his crime. Some of our subjects are aware of the power of suggestion, and when absolutely resolved to commit an act for which they fear that their courage or audacity may fail when the moment arrives, they take the precaution of receiving the suggestion from their companions.

The danger of these criminal suggestions is increased by the fact that at the will of the experimenter, the act may be accomplished several hours, and even several days, after the date of suggestion. Facts of this kind, which were first reported by Richet, are not exceptional, and have been repeatedly observed by us.

The reality of this class of facts cannot now be disputed, but the difficulty of proof in any given case is considerable. We have not, in the case of impulsive acts, the same objective criterion as we have in hallucinations and in the paralysis of movements and of sensation. It is, therefore, necessary for the expert to be cautious in his judgment.

Loss of memory is one chief characteristic of the facts of suggestion. The hypnotic subject does not know from whom, when, and how the suggestion was received. This amnesia may be either spontaneous or suggested,

17

and it is a phenomenon of the waking state, which disappears when the subject is hypnotized anew. The recollection of all which occurred during hypnosis is then revived, and the subject is able to indicate, often with remarkable precision, the author of the suggestion, the place, day, and hour when it was made to him, always supposing that he has received no special suggestion of complete oblivion. Hence the question occurs whether an accused person who appeals to an hypnotic suggestion for his defence, and who submits to experiment, can be profitably examined at a time when he displays all the physical characteristics peculiar to the somnambulist state, so that there is no danger of imposture. We have had occasion to show that some subjects are in this state capable of suppressing the truth, and Pitres has shown that deceit was not impossible. An hypnotic subject may at the same time be criminal, and suggestion must be accepted only so far as it admits of material proof, or at any rate as far as it can be necessarily deduced from the facts of the case.

Simulation is not the only danger to be avoided in the examination of a somnambulist subject. It is possible that a magistrate or physician may, by the persistence of his questions and his authoritative voice, unconsciously give suggestions which modify the subject's recollections, and give rise to hallucinations of memory. There is a further danger that the examination of the subject may conflict with a previous suggestion, by which he had been forbidden to speak of certain events. It is true that a little dexterity will overcome this defence, as, for instance, by assuming, by means of suggestion, the personality of the first experimenter. But the facts we have given are enough to show that the examination of an hypnotic subject does not afford a sufficient warrant of her sincerity.

There is still more reason for condemning an inquiry by means of hypnotism. It has been suggested that a suspected or accused person might be hypnotized against his will, in order to obtain from him admissions or information respecting the facts of the accusation. This process, which resembles that of torture, would have the same danger of leading a suspected person to confess crimes of which he was not really guilty.

We should say, in conclusion, that the suggestion of crime, whether made in or out of the hypnotic state, upon hypnotic or neuropathic subjects, can only be proved by the physical characteristics furnished by the subject. The medical expert, whose duty it is to bring justice to light, and not to find victims for the law, must be content with this observation. He may prove by experiment that a given subject is or is not susceptible to hypnotism, and that the phenomena in question may be produced during hypnosis, or under the influence of an hypnotic suggestion, but he can only give evidence of the possibility of the fact. It is for others to decide whether the fact really occurred.

When once the suggestion has been proved, it remains to be seen what is the penal and moral responsibility of the individual who has acted under the influence of an hypnotic suggestion. We think that some authors have been too hasty in deciding that the moral responsibility of the subject has absolutely ceased. We should at any rate inquire why he consented to be hypnotized, unless this was done suddenly, or by force or guile. Even if he was not aware of the experimenter's purpose in hypnotizing him, he must be held responsible for having voluntarily alienated his free will. This principle may certainly be established, and it is still more applicable if the subject was aware before he was hypnotized for what criminal act it was proposed to employ him. This hypothesis is by no means impossible.

It is possible that a subject susceptible to hypnotism might be found in a band of swindlers or murderers who would willingly become the recipient of criminal suggestions. We can readily understand the use of suggestion in such circumstances, since those who act under the influence of hypnotic suggestion display more daring and courage, and even more intelligence, than when they act from their own impulse.

Even when the subject has been hypnotized without his consent, when he has been taken by surprise, and has received the suggestion during his waking state, and has consequently incurred no moral responsibility, it cannot be denied that society is justified in defending itself against such a dangerous subject. So far, hypnotism has only made a casual appearance in a court of justice. But this state of things might change at any time, and suggestion may become an instrument of criminal practices. No one who studies the history of crime will maintain that society ought not to guard against such a danger. Hypnotic criminals ought to be treated like insane criminals.*

Since the possibility of curing a certain number of nervous diseases by means of hypnotism is established,

* We are glad to see that Tarde shares this opinion. See Alcan, *La criminalité comparée*, p. 142.

it cannot be disputed that physicians are justified in making use of it, under the same reservation as any other methods of therapeutics. The physician's responsibility is diminished if he has to treat an affection which would not yield to other measures; if he has obtained the consent of his patient and the concurrence of the patient's friends, and, finally; if he can show that he has acted prudently, with due consideration of the danger incurred by the patient, and with proper precautions against these risks.

Some reservations must be made with respect to experiment, strictly so called.

Some facts recently observed in Italy seem to indicate that the practice of hypnotism may produce in the subject permanent nervous affections. Yet to assert that experiments must, therefore, be prohibited would imply that there are truths which it is better not to know. It is, however, certain that the prejudice should be overcome by those who have excited it, and that men are not freed from all responsibility by the fact that they possess a diploma. If, instead of considering the interest of the patient, the result of the experiment is the only object in view, it may come to pass that, on pretence of shedding light on the highest problems of physiology or psychology, inquirers who are not thoroughly acquainted with anatomy, physiology, and therapeutics, may imperil the lives of subjects committed to their care. Ignorance of the danger is no valid excuse for the imprudence of the experimenter.

It is only on these conditions that experiments on the human subject should be performed, and it may be said in passing that there is nothing in them which need appear shocking. It is indeed an every-day practice, since in laboratories and hospitals, patients or students are always willing to submit to the action of drugs or to physiological experiments, and experiments in hypnotism may be performed without inconvenience under the same conditions and safeguards.

With respect to the performance of such experiments in public, it should be condemned, just as we condemn public dissections of the dead body, and vivisection in public. It is certain that there are still graver objections to hypnotic exhibitions, since they are liable to produce nervous affections, even in those who do not propose to be the subjects of experiment.

THE END.